Understanding Mineralogy

Understanding Mineralogy

Cyaden Ross

MURPHY & MOORE
www.murphy-moorepublishing.com

Published by Murphy & Moore Publishing,
1 Rockefeller Plaza,
New York City, NY 10020, USA
www.murphy-moorepublishing.com

Understanding Mineralogy
Cyaden Ross

International Standard Book Number: 978-1-63987-557-3 (Hardback)

Cataloging-in-Publication Data

Understanding mineralogy / Cyaden Ross.
p. cm.
Includes bibliographical references and index.
ISBN 978-1-63987-557-3
1. Mineralogy. 2. Minerals. 3. Physical geology. 4. Earth sciences.
I. Ross, Cyaden.
QE363.2 .T49 2022
549--dc23

TABLE OF CONTENTS

PREFACE

The sub branch of geology which focuses on analyzing the crystal structure, physical characteristics and chemical properties of various minerals and mineralized artifacts is known as mineralogy. This includes the study of processes involved in the formation, classification, geographical distribution and utilization of minerals. The discipline classifies minerals into various categories such as carbonates, sulfates, phosphates, silicates, oxides, halides, etc. These are studied using a wide variety of techniques like X-ray diffraction, photon induced X-ray emission spectroscopy, scanning electron microscopy, ultraviolet spectroscopy and X-ray fluorescence. Some of the key processes which are studied within this field are sublimation from volcanic gases, deposition of aqueous solutions, crystallization of lava and recrystallization. This book is a valuable compilation of topics, ranging from the basic to the most complex theories and principles in the field of mineralogy. It aims to shed light on some of the unexplored aspects of mineralogy. Those in search of information to further their knowledge will be greatly assisted by this book.

Given below is the chapter wise description of the book:

Chapter 1- Mineralogy is the branch of geology concerned with the study of minerals. It is the scientific study of the chemistry, crystal structure, physical properties, occurrence and distribution, and the origins of minerals and mineralized artifacts. This chapter introduces mineralogy and its related aspects in an easy to understand manner.

Chapter 2- Minerals are naturally occurring solids with a definite but not necessarily fixed chemical composition. Years of weathering and erosion exposes them to the elements of nature which interact and modify their chemical compositions. This chapter deals with mineral chemistry and its related aspects like metamorphism and metasomatism in detail.

Chapter 3- Mineral physics is the science of geological and geo-inspired materials with emphasis on their structures, properties and behavior. Optical properties, magnetic properties, mineral strength, electrical properties, etc. are aspects which fall under its domain. This chapter closely examines all these concepts associated with mineral physics to provide an extensive understanding of the subject.

Chapter 4- Minerals are classified according to their chemical properties, with the anion being the chemical basis for classification. Silicate minerals, native element minerals, oxide minerals, halide minerals, sulfate minerals and phosphate minerals are the classes of minerals that are covered in this chapter in a clear and concise manner.

Chapter 5- Gemstones are minerals which are highly prized for their beauty, durability and rarity. Gemstones, after being cut and polished, can be used in jewelry. Gemology is the science of studying, cutting and valuing of precious stones regardless of whether they are artificial, imitation or natural. This chapter focuses on gemstones and gemology and will help in obtaining extensive understanding of the subject matter.

Chapter 6- Crystallography is the study of distinguishing the arrangement and bonding of atoms in crystalline solids and with the geometric structure of crystal lattices. Most minerals occur naturally as crystals and thus are studied under crystallography. Crystal symmetry, crystal systems, crystal classes, crystal lattices, etc. are some of its aspects. This chapter has been carefully written to provide an easy understanding of the subject matter and its related points.

Indeed, my job was extremely crucial and challenging as I had to ensure that every chapter is informative and structured in a student-friendly manner. I am thankful for the support provided by my family and colleagues during the completion of this book.

Cyaden Ross

Chapter 1

Mineralogy: An Introduction

Mineralogy is the branch of geology concerned with the study of minerals. It is the scientific study of the chemistry, crystal structure, physical properties, occurrence and distribution, and the origins of minerals and mineralized artifacts. This chapter introduces mineralogy and its related aspects in an easy to understand manner.

The science of mineralogy is a branch of the earth sciences that is concerned with studying minerals and their physical and chemical properties. Within mineralogy there are also those who study how minerals are formed, where they are geographically located, as well as their potential uses. Like many sciences, mineralogy has its origins in several ancient civilizations, and it has been concerned primarily with the various methods of classification of minerals for most of its history. Modern-day mineralogy has been expanded by advances in other sciences, such as biology and chemistry, to shed even more light on the nature of the materials that form the earth we live on.

The ancient Greek philosopher Aristotle was one of the first people to theorize extensively about the origins and properties of minerals. His ideas were new and advanced for the time, but he and his contemporaries were largely incorrect in their assumptions. For example, it was a widely held belief in ancient Greece that the mineral asbestos was a kind of vegetable. Nevertheless, these ancient theories provided a starting point for the evolution of mineralogy as we have come to know it. It was not until the 16th century that mineralogy began to take a form that is recognizable to us, largely thanks to the work of German scientist Georgius Agricola.

For example, it was a widely held belief in ancient Greece that the mineral asbestos was a kind of vegetable. Nevertheless, these ancient theories provided a starting point for the evolution of mineralogy as we have come to know it. It was not until the 16th century that mineralogy began to take a form that is recognizable to us, largely thanks to the work of German scientist Georgius Agricola.

A mineral is a naturally-occurring, homogeneous solid with a definite, but generally no fixed, chemical composition and an ordered atomic arrangement. It is usually formed inorganic processes.

Let's look at the five parts of this definition:

- "Naturally occurring" means that synthetic compounds not known to occur in natur cannot have a mineral name. However, it may occur anywhere, other planets, deep in the earth, as long as there exists a natural sample to describe.
- "Homogeneous solid" means that it must be chemically and physically homogeneous down to the basic repeat unit of the atoms. It will then have absolutely predictabl physical properties

(density, compressibility, index of refraction, etc.). This means tha rocks such as granite or basalt are not minerals because they contain more than on compound.

- "Definite, but generally not fixed, composition" means that atoms, or groups of atom must occur in specific ratios. For ionic crystals (i.e. most minerals) ratios of cations two anions will be constrained by charge balance, however, atoms of similar charge an ionic radius may substitute freely for one another; hence definite, but not fixed.
- "Ordered atomic arrangement" means crystalline. Crystalline materials are three dimensional periodic arrays of precise geometric arrangement of atoms. Glasses such an obsidian, which are disordered solids, liquids (e.g., water, mercury), and gases (e.g., air are not minerals).
- "Inorganic processes" means that crystalline organic compounds formed by organism are generally not considered minerals. However, carbonate shells are minerals because they are identical to compounds formed by purely inorganic processes.

An abbreviated definition of a mineral would be "a natural, crystalline phase". Chemist has a precise definition of a phase. A phase is that part of a system which is physically a chemically homogeneous within itself and is surrounded by a boundary such that it is mechanically separable from the rest of the system. The third part of our definition of mineral leads us to a brief discussion of stoichiometry, the ratios in which different element (atoms) occur in minerals. Because minerals are crystals, dissimilar elements must occur in fixed ratios to one another. However, complete free substitution of very similar element (e.g., Mg^{+2} and Fe^{+2} which are very similar in charge (valence) and radius is very commo and usually results in a crystalline solution (solid solution). For example, the mineral forsterite (Mg_2SiO_4) and fayalite (Fe_2SiO_4) are members of the olivine group and have the same crystal structure, that is, the same geometric arrangement of atoms. Mg and Fe substitute freely for each other in this structure, and all compositions between the two extremes, forsterite and fayalite, may occur. However, Mg or Fe do not substitute for Si or O, so that the three components, Mg/Fe, Si and O always maintain the same 2 to 1 to 4 ratio because the ratio is fixed by the crystalline structure. These two minerals are called endmembers of the olivine series and represent extremes or "pure" compositions. Because these two minerals have the same structure, they are called isomorphs and the series, an isomorphous series.

In contrast to the isomorphous series, it is also common for a single compound (composition) to occur with different crystal structures. Each of these structures is then a different mineral and, in general, will be stable under different conditions of temperature and pressure. Different structural modifications of the same compound are called polymorphs. An example of polymorphism is the different minerals of SiO_2 (silica); alphaquartz, beta-quartz, tridymite, cristobalite, coesite, and stishovite. Although each of these has the same formula and composition, they are different minerals because they have different crystal structures. Each is stable under a different set of temperature and pressure conditions, and the presence of one of these in a rock may be used to infer the conditions of formation of a rock. Another familiar example of polymorphism is graphite and diamond, two different minerals with the same formula, C (carbon).

Glasses (obsidian), liquids, and gases however, are not crystalline, and the elements in them may occur in any ratios, so they are not minerals. So in order for a natural compound to be a mineral, it must have a unique composition and structure.

Composition of the Earth's Crust

The earth's crust is composed of many kinds of rocks, each of which is an aggregate of one or more minerals. In geology, the term mineral describes any naturally-occurring solid substance with a specific composition and crystal structure. A mineral's composition refers to the kinds and proportions of elements making up the mineral. The way these elements are packed together determines the structure of the mineral. More than 3,500 different minerals have been identified. There are only 12 common elements (oxygen, silicon, aluminum, iron, calcium, sodium, potassium, magnesium, titanium, hydrogen, manganese, phosphorus) that occur in the earth's crust. All other naturally occurring elements are found in very minor or trace amounts. Silicon and oxygen are the most abundant crustal elements, together comprising more than 70 percent by weight it is therefore not surprising that the most abundant crustal minerals are the silicates, followed by the oxides. Other important types of minerals include: the carbonates the sulfides (e.g. galena, PbS) and the sulfates. Most of the abundant minerals in the earth's crust are not of commercial value. Economically valuable minerals (metallic and nonmetallic) that provide the raw materials for industry tend to be rare and hard to find. Therefore, considerable effort and skill is necessary for finding where they occur and extracting them in sufficient quantities. Table shows the elemental chemical composition of the Earth's crust in order of abundance.

Table: The elements in the Earth's crust.

Element name	Symbol	Percentage by weight of the Earth's crust
Oxygen	O	46,6
Silicon	Si	27,7
Aluminium	Al	8,1
Iron	Fe	5,0
Calcium	Ca	3.6
Sodium	Na	2,8
Potassium	K	2,6
Magnesium	Mg	2,1
All other elements		1,5

The table above shows the elemental chemical composition of the Earth's crust. They will vary depending on the way they were calculated and the source. 98.5% of the Earth's crust consists of oxygen, silicon, aluminum, iron, calcium, sodium, potassium and magnesium. All other elements account for approximately 1.5% of the volume of the Earth's crust.

Characteristics of Minerals

The physical properties of a mineral are determined by its chemical composition and its crystalline structure. Within the limits of the permissible variation in chemical composition, different samples of a single mineral species are expected to display the same set of physical properties. These characteristic physical properties are therefore very useful to the field geologist in identifying and describing a specimen.

Properties which describe the physical appearance of a mineral specimen include color, streak, and luster. Mass-dependent properties include density; mechanical properties include hardness, cleavage, fracture, and tenacity. Properties relating to the growth patterns and physical appearance of crystals, both individually and in aggregate are described in terms of crystal habit, crystal form, and crystal system:

- Crystal form and habit (shape).
- Luster and transparency.
- Color and streak.
- Cleavage, fracture, and parting.
- Tenacity.
- Density.
- Hardness.

Crystal Form and Habit

The crystal faces developed on a specimen may arise either as a result of growth or of cleavage. In either case, they reflect the internal symmetry of the crystal structure that makes the mineral unique. The crystal faces commonly seen on quartz are growth faces and represent the slowest growing directions in the structure. Quartz grows rapidly along its caxis (tree-fold or trigonal symmetry axis) direction and so never shows faces perpendicular to this direction. On the other hand, calcite rhomb faces and mica plates are cleavages and represent the weakest chemical bonds in the structure. There is a complex terminology for crystal faces, but some obvious names for faces are prisms and pyramids. A prism is a face that is perpendicular to a major axis of the crystal, whereas a pyramid is one that is not perpendicular to any major axis.

Crystals that commonly develop prism faces are said to have a prismatic or columnar habit. Crystals that grow in fine needles are acicular; crystals growing flat plates are tabular. Crystals forming radiating sprays of needles or fibers are stellate. Crystals forming parallel fibers are fibrous, and crystals forming branching, tree-like growths are dendritic.

Luster and Transparency

The way a mineral transmits or reflects light is a diagnostic property. The transparency may be opaque, translucent, or transparent. This reflectance property is called luster. Native metals and many sulfides are opaque and reflect most of the light hitting their surfaces and have a metallic luster. Other opaque or nearly opaque oxides may appear dull, or resinous. Transparent minerals with a high index of refraction such as diamond appear brilliant and are said to have an adamantine luster, whereas those with a lower index of refraction such as quartz or calcite appear glassy and are said to have a vitreous luster.

Color and Streak

Color is fairly self-explanatory property describing the reflectance. Metallic minerals are white, gray, or yellow. The presence of transition metals with unfilled electron shells (e.g. V, Cr, Mn, Fe,

Co, Ni, and Cu) in oxide and silicate minerals causes them to be opaque or strongly colored so that the streak, the mark that they leave when scratched on a white ceramic tile, will also be strongly colored.

Cleavage, Fracture and Parting

Because bonding is not of equal strength in all directions in most crystals, they will tend to break along crystallographic directions giving them a fracture property that reflects the underlying structure and is frequently diagnostic. A perfect cleavage results in regular flat faces resembling growth faces such as in mica, or calcite. A less well developed cleavage is said to be imperfect, or if very weak, a parting. If a fracture is irregular and results in a rough surface, it is hackly. If the irregular fracture propagates as a single surface resulting in a shiny surface as in glass, the fracture is said to be conchoidal.

Tenacity

Tenacity is the ability of a mineral to deform plastically under stress. Minerals may be brittle, that is, they do not deform, but rather fracture, under stress as do most silicates and oxides. They may be sectile, or be able to deform so that they can be cut with a knife. Or, they may be ductile and deform readily under stress as does gold.

Density

Density is a well-defined physical property measured in g/cm^3. Most silicates of light element have densities in the range 2.6 to 3.5. Sulfides are typically 5 to 6. Iron metal about 8, lead about 13, gold about 19, and osmium, the densest substance, and a native element mineral is 22.

Hardness

Hardness is usually tested by seeing if some standard minerals are able to scratch others. A standard scale was developed by Friedrich Mohs in 1812. The standard minerals making up the Mohs scale of hardness are:

- Talc,
- Gypsum,
- Calcite,
- Fluorite,
- Apatite,
- Orthoclase,
- Quartz,
- Topaz,
- Corundum,
- Diamond.

This scale is approximately linear up to corundum, but diamond is approximately 5 times harder than corundum.

Unique Properties

A few minerals may have easily tested unique properties that may greatly aid identification. For example, halite (NaCl) (common table salt) and sylvite (KCl) are very similar in most of their physical properties, but have a distinctly different taste on the tongue, with sylvite having a more bitter taste. Another unique property that can be used to distinguish between otherwise similar back opaque minerals is magnetism. For example, magnetite (Fe_3O_4), ilmenite ($FeTiO_3$), and pyrolusite (M_nO_2) are all dense, black, opaque minerals which can easily be distinguished by testing the magnetism with a magnet. Magnetite is strongly magnetic and can be permanently magnetized to form a lodestone; ilmenite is weakly magnetic; and pyrolusite is not magnetic at all.

Other Properties

There are numerous other properties that are diagnostic of minerals, but which generally require more sophisticated devices to measure or detect. For example, minerals containing the elements U or Th are radioactive, and this radioactivity can be easily detected with a Geiger counter. Examples of radioactive minerals are uraninite (UO_2), thorite ($ThSiO_4$), and carnotite ($K_2 (UO_2) (VO_4)_2 rH_2O$). Some minerals may also be fluorescent under ultraviolet light that is they absorb UV lighta and emit in the visible. Other optical properties such as index of refraction and pleochroism (differential light absorption) require an optical microscope to measure. Electrical conductivity is an important physical property but requires an impedance bridge to measure. In general native metals are good conductors; sulfides of transition metals are semi-conductors, whereas most oxygen-bearing minerals (i.e., silicates, carbonates, oxides, etc.) are insulators. Additionally, quartz (SiO_2) is piezoelectric (develops an electrical charge at opposite end under an applied mechanical stress); and tourmaline is pyroelectric (develops an electrical charge at opposite end under an applied thermal gradient).

Branches of Mineralogy

Mineralogy is the branch of geology concerned with the study of minerals. A mineral is a naturally occurring, homogeneous solid with a definite chemical composition and a highly ordered atomic structure. A homogeneous substance is one that can be divided into repeating units that are exactly the same. A mineral, by definition, cannot be a liquid or a gas. The chemical composition of a mineral is definite, meaning a particular mineral is always composed of the same ratio of elements, and this composition can be shown using a chemical formula. The atoms in a mineral are arranged in a highly ordered fashion, called a crystal lattice structure.

Minerals have been an important part of our society since the time of prehistoric man. Early humans carved tools out of minerals such as quartz. Pottery has been made of various clays since ancient times. Sodium chloride, also known as the mineral halite, has been used in food preservation techniques for millions of years. Mining of useful minerals out of ores became widespread hundreds of years ago, a practice still in use today.

Crystallography

There are several different branches of mineralogy. Mineralogists can focus on very specific studies, from crystal structure to classification or chemical composition. Crystallography, for example, is the study of the crystal lattice structure of minerals. The atoms in a mineral are arranged in a highly ordered fashion. This ordered arrangement produces crystals of definite size and shape. A particular mineral sample is made up of repeating crystal units. Each crystal that makes up the mineral has the same shape. There are six basic shapes a mineral crystal can have. The shape of the crystal, as well as how tightly packed the atoms are in the crystal, help determine the physical properties of the mineral. Crystals that are allowed to grow with plenty of open space will form nearly perfect structures, and those that form in more cramped conditions will display imperfections in the crystal shape.

Crystal and Conformational Chemistry

Crystal chemistry is the branch of mineralogy that deals with how the chemical composition of a mineral relates to its crystal structure. The chemical bonds formed between atoms determine the crystal shape as well as the chemical and physical properties of the mineral. There are three different types of chemical bonds present in minerals—ionic, covalent, and metallic. In ionic bonding, an atom with a positive charge binds to an atom with a negative charge through electrostatic attraction. Minerals with ionic bonds tend to be poor conductors of heat and electricity, have low melting points, and are brittle. Halite and fluorite are both minerals formed by ionic bonds. In covalent bonding, electrons are shared between two atoms. This type of bonding is stronger than ionic bonding, which means minerals with covalent bonds have higher melting points and are harder than those with ionic bonds. These minerals are also poor conductors of heat and electricity and are brittle. Examples of covalently bonded minerals include quartz and diamond. Metallic bonding occurs between atoms of metals. In this type of bond, the outer electrons of the atom are free to move, and are shared between all of the other atoms in the substance. This special structure is the reason metals are good conductors of heat and electricity, are malleable, soft, and have lower melting points. Copper, silver, and gold are all minerals formed by metallic bonding.

Physical Mineralogy

Physical mineralogy is concerned with the physical properties and descriptions of minerals. Minerals can be described using several physical attributes, including hardness, specific gravity, luster, color, streak, and cleavage.

The hardness of a mineral can be determined by a scratch test. The scratch test establishes how easily a mark can be made on a mineral sample using different materials. If a mark is made easily, the mineral is not very hard. If no mark can be made, then the mineral is quite hard. The hardness is then measured on a scale of 1-10, called Mohs' hardness scale, named after the Austrian scientist F. Mohs, who developed this procedure. If a fingernail can scratch a particular mineral, it would have a hardness of 2.5. If a penny can scratch it, its hardness is around 3. If a mineral can be scratched by glass, its hardness is 5.5. If it can be scratched by unglazed porcelain, it has hardness between 6 and 6.5, and if a steel file can leave a mark, it has a hardness of 6-7. Talc is the softest mineral with a hardness rating of 1, while diamond is the hardest, rated 10. The specific gravity of a mineral is the ratio of the mass of a particular volume of the mineral to that of the same volume of water. All minerals have a specific gravity greater than 1.

The luster of a mineral is the appearance of its surface when light is reflected off of it. Minerals can have metallic or nonmetallic luster. Minerals with metallic luster look shiny like a metal. Nonmetallic minerals can have various appearances, such as vitreous (glassy), greasy, silky, brilliant (like a diamond), or pearly.

The color of a mineral sample cannot be used to definitively identify the mineral because of impurities that may be present; however, the color can narrow down the identity of a mineral to a few choices. The streak of a mineral is the color of its powdered form. Rubbing the mineral across unglazed porcelain square, called a streak plate, can best show streak color. A mineral will have a characteristic streak color, although more than one mineral may have the same color. Therefore, streak is not a definitive identification tool, although it may be used to verify the identity of a mineral of suspected composition.

A mineral exhibits cleavage when it breaks along a certain direction or plane, producing a flat surface along the break. When a mineral shatters, rather than breaks along planes, it exhibits fracture. Cleavage is characteristic of particular minerals such as feldspar, while minerals such as quartz show fracture. Each of these physical properties can be used to determine the chemical identity of an unknown mineral, and together are the focus of the branch of mineralogy called physical mineralogy.

Other Branches

Descriptive mineralogists use the properties discussed in physical mineralogy to name and classify new minerals. Determinative mineralogy is the branch of mineralogy that deals with identifying unknown minerals, also using the physical properties of minerals. Other branches of mineralogy include chemical mineralogy (identifying minerals to determine the chemical composition of the earth's crust), optical mineralogy (using light to determine the crystal structure of minerals), x-ray mineralogy (using x-ray diffraction techniques to determine the crystal structure of minerals), and economic mineralogy (the study of new, economically important uses for minerals). All of the branches of mineralogy together describe the physical and chemical properties of minerals and their uses.

Mineralogy is an important discipline for several reasons. For one, the study of the composition of the earth's crust gives scientists an idea of how Earth was formed. The discovery of new minerals could provide useful materials for industry. The study of the chemical properties of minerals could lead to the discovery of new uses for Earth's mineral resources. Mining ores for their mineral components provides the materials for lasers, buildings, and jewelry. Each of the branches of mineralogy contributes to the indispensable knowledge base of minerals and their uses.

Ore Mineralogy

Mining, processing, and refining of minerals accounts for a significant portion of global GDP. Characterization of the mineral, or in most cases minerals, making up an ore body (ore minerals) is critical for their efficient and sustainable exploitation and for the validation and refinement of genetic models that can underpin exploration, in turn leading to new discoveries to satisfy future demand. Ore minerals, in the context of the ore deposit that hosts them, can give important insights into fundamental topics of solid state chemistry and ore-forming processes.

Definitions of what constitutes an ore mineral vary significantly from source to source. Broadly, and in the context used in key reference texts, ore minerals are those containing one or more elements of economic interest, as well as related, co-existing mineral phases, which were concentrated and deposited via the same physical-chemical processes. Many ore minerals, notably the economically important sulphide, and oxide groups, are opaque, and are traditionally studied in reflected rather than transmitted light, a quality distinguishing them from rock-forming minerals. There are, however, many examples of minerals that represent an economic resource, but which are not opaque, and likewise, many opaque minerals that represent waste in many mining operations. A good example is hematite from iron oxide-copper-gold (IOCG) deposits, such as Olympic Dam, South Australia. This contribution therefore considers ore minerals in a loose sense, as those minerals concentrated within mineral deposits, whether of economic value or not.

The concept of "ore minerals" has changed over the past generation, as mines have become larger and in general of a lower grade, and as new discoveries of giant, high-grade mineral deposits have become less frequent relative to exploration dollars spent. Moreover, the global transition to new technologies has modified the patterns of demand for individual commodities. The emergence of the critical metals sector, for example, has catalyzed both exploration and research interest into a range of new mineral resources that were previously regarded as sub-economic or waste. Examples include minerals that host lithium (Li), scandium (Sc), or the rare earth elements (REE). Such commodities may occur in traditional gangue minerals (e.g., silicates, carbonates etc.), or as trace substituents within sulphides, as in the case of indium (In) or germanium (Ge), for example. The boundaries of ore mineralogy are further blurred by recognition among researchers that all of the minerals making up ore assemblages are best addressed collectively, and that gangue and accessory components within the ores, rather than the ore minerals themselves, may provide valuable information on ore genesis, formation conditions, or serve as indicators in mineral exploration.

Advances

Microanalytical Methods

The scanning electron microscope (SEM) and electron probe microanalyzer (EPMA) were two of the most important developments in mineralogy in the period 1950–1980. In mineral deposit geology, direct measurement of ore-forming fluids that are preserved within fluid inclusions trapped within minerals, and stable isotopes, stand out as the landmark technologies, which, during that same period, changed the way in which ores and ore minerals are studied. Analytical strategies for the routine study of ore minerals have undergone a revolution during the past 30 or so years due to an explosion of different, complementary in-situ microanalytical capabilities. The rapid expansion of microanalysis, with improved resolution and sensitivity, enables the ore mineralogist to observe measure and quantify in ways that were not previously possible.

Laser-Ablation Inductively-Coupled Plasma Mass Spectrometry (LA-ICP-MS)

LA-ICP-MS has rapidly established itself as a requisite research tool with application across the earth sciences. For ore mineralogists, LA-ICP-MS represents a rapid and relatively inexpensive method for accurate, precise in-situ measurement of concentrations of trace elements in mineral matrices, and to map the distributions of those elements at the grain-scale. This, in turn, enables genetic models for mineral deposits to be refined. Large LA-ICP-MS datasets can form a foundation for understanding trace element incorporation and release from ore minerals, as well as for the development of geochemical vectors of value in mineral exploration. Coupled with automated mineralogy, LA-ICP-MS allows for reliable deportment models to be generated for elements of interest. Moreover, LA-ICP-MS also offers a platform for in-situ geochronology, and in-situ chemical characterization of fluid inclusions. Cook et al. provide an overview of current applications of LA-ICP-MS to the trace element analysis of minerals in hydrothermal ore deposits, also noting research gaps, some of the outstanding issues, and future opportunities. The rapidly expanding capabilities of current LA-ICP-MS instrumentation, including both quadruple and multi-collector systems, include in-situ determination of stable isotopes, and grain-scale mapping of isotopic signatures. Outstanding challenges include inadequately constrained issues of down-hole fractionation, the impact of matrix effects, and concerns about how the behavior of certain minerals during ablation may induce fractionation. To this, we add the long-standing question of how LA-ICP-MS data from fine-grained, heterogeneous, complex ores can best be interpreted, when the sample volumes analyzed by LA-ICP-MS contain measured elements that are in solid solution within the host, as well as sub-micron-sized mineral inclusions, which, in many cases, are just a few nanometers across.

Figure: Back-scatter electron (BSE) image of compositionally-zoned hematite. Brighter bands are enriched in U, W, Mo, and Sn relative to darker bands.

Focus Ion Beam-Scanning Electron Microscopy (FIB-SEM)

FIB-SEM methods have helped resolve problems that are associated with in-situ extraction of small amounts of sample material that has been well-characterised at the micron-scale, e.g., by EPMA or LA-ICP-MS, from the surface of a polished mount for milling and subsequent analysis by transmission electron microscopy. Direct application of FIB-SEM methods in ore mineralogy is expanding steadily. The technique has revolutionized mineral characterization by enabling scales of observation to be bridged, i.e., linking observations at the nanometre-scale to features observable at the scale of microns in reflected light or by SEM. In tomographic analysis mode, the FIB-SEM provides three-dimensional visualisation of a sample with resolution well below the micron-scale.

Figure illustrates an example in which FIB-SEM has been used to visualise and extract thin foils in-situ from iron-oxides to examine and identify nanoscale inclusions using high-angle angular dark field scanning transmission electron microscopy (HAADF STEM). The magnetite illustrated occurs as a magmatic accessory in the host granite (Roxby Downs Granite; RDG) to the Olympic Dam Cu-U-Au-Ag deposit. In general, magmatic magnetite in the RDG is associated with ilmenite, apatite and zircon, as well as minor titanite. The binary inclusions that are shown here were observed at the micron scale only for a sub-population of magnetite grains. Nanoscale investigation shows they are however ubiquitously present throughout the magnetite although they vary in size down to the smallest nanoparticles. The presence of such inclusions in RDG magnetite is taken as further evidence for mixing between granitic and mafic magmas, as was suggested by studies of apatite that contain pyrrhotite-fluorite inclusions, relationships among feldspars in the RDG and indeed observations on intrusive rocks from other granitic plutons elsewhere.

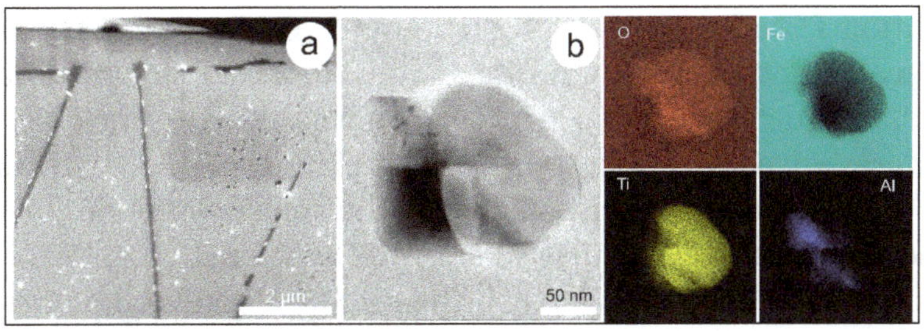

Figure shows (a) Indicates that Secondary electron (SE) image of a FIB-SEM-prepared TEM foil con-sisting of magnetite with dense fields of sub-µm-scale, binary inclusions of ilmenite and hercynite. Ilmenite is also present as trellis-like exsolution (darker grey). Colour on the SE image is not indicative of the average atomic mass of the mineral species and their identification requires dedicated nanoscale analysis; (b) HAADF STEM image of one of the larger composite inclusions and transmission electron microscope—energy-dispersive X-ray spectrometry (EDS) element maps of Fe, O, Ti and Al clearly showing two components (ilmenite and hercynite).

HAADF STEM Imaging and STEM-EDS Mapping

Transmission electron microscopy has been a routine mineralogical technique dating back to the middle of the 20th Century, but continues to advance at an extraordinary rate. The application of HAADF STEM is, however, a more recent development. HAADF STEM enhances Z-scale contrast, enabling atomic-scale resolution with applicability to many minerals that are found within ore deposits, e.g., mixed-layer chalcogenide compounds of the tetradymite group.

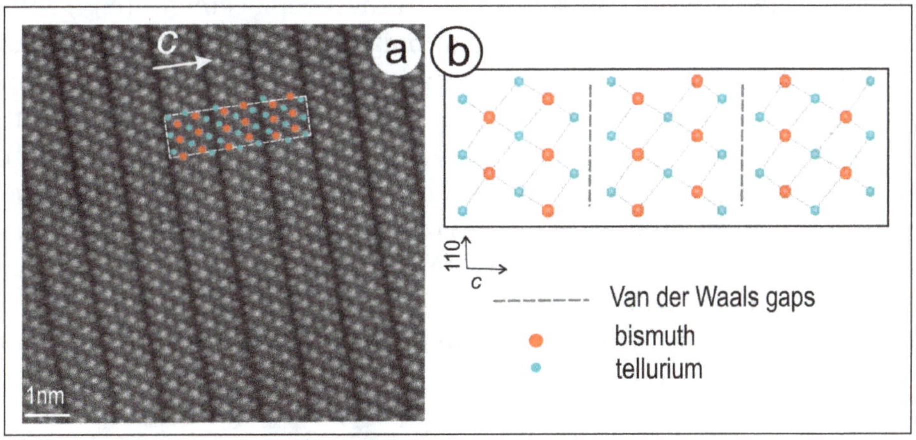

Figure shows (a) Indicates that High-resolution HAADF STEM image of tellurobismuthite, Bi2Te3; and (b) simulated crystal structure viewed down the axis). Match between the marked atoms, i.e., Bi—brighter and Te—darker; in (a) with the structure in (b). Dashed line shows the size of the unit cell. The three layer-subunits containing the characteristic atom array for tellurobismuthite, Te-Bi-Te-Bi-Te, are separated by darker lines corresponding to Van der Waals gaps.

The atomic-scale resolution possible with contemporary state-of-the-art STEM instrumentation in HAADF STEM mode is particularly well-suited to visualise the structures of complex sulphosalts e.g., Pb-Bi-sulphosalts, Zn-bearing smectites or disordered intergrowths among mixed-layer compounds, such as the REE-fluorocarbonates. Atomic-scale HAADF STEM imaging is a more direct tool than high-resolution TEM since it provides a better control on the correlation between crystal structures and chemical variation. It is therefore an ideal tool for resolving issues of polysomatism within mineral series, in which the crystal structural motifs or layer stacking change gradually. In combination with energy-dispersive X-ray (EDX) mapping, the HAADF STEM technique is well suited to the identification of nanoparticles, and even the smallest atomic clusters that are present within a mineral. The technique is particularly useful for heavier (U, Au, Te, Bi, Pb, REE etc. rather than lighter atoms. Remarkable insights have, however, been obtained by HAADF STEM into the nanoscale mechanisms responsible for precipitation of siliceous magnetite. Electron energy loss spectroscopy (EELS) on a TEM platform is also an important technique that can be used to show the oxidation state of elements of interest within a nanoparticulate mineral and any variation in such oxidation state at the nanoscale, e.g. Based on our accumulated experience, HAADF STEM imaging works better on FIB-prepared foils (often with variable thickness and distortions) than HR TEM imaging. Future research topics with potential in ore mineralogy include: (i) trace element incorporation in minerals and their release; (ii) polysomatism in mixed-layer compounds; and, (iii) lattice-scale order/disorder as a tool to constrain equilibrium crystallization versus mineral replacement.

To illustrate an application of HAADF STEM, an early hydrothermal association of hyalophane-albite within the core of a magmatic K-feldspar phenocryst (mm-sized), with rapakivi texture from granite outside the orebody at Olympic Dam. The sample is considered "fresh", based on whole rock geochemical data analysis and the preservation of magmatic amphibole and biotite. In IOCG deposits, alteration is often defined based on mineral associations that are constrained regionally or at the scale of the deposit. This example shows, however, that intensive hydrothermal alteration with the development of calc-silicates and their replacement. (e.g., chlorite replacing epidote) can occur at the nanoscale, even though magmatic features (e.g., cryptoperthites) are preserved. The assemblage also contains abundant, nm-scale inclusions of hematite and larger grains of sericite epitaxially intergrown with chlorite. Such associations draw attention to the telescoping character of alteration stages in IOCG deposits, which are otherwise treated only as well-defined spatial-temporal stages. Moreover, these mineral reactions are driven by coupled dissolution-reprecipitation reaction (CDRR), and thus chemical exchange is trapped within areas ranging in scales from a few nm up to several meters. Considering that we show here a representative case for an IOCG deposit with high metal endowment, it may be that these types of reaction are paramount in contributing to the fertility of an IOCG system by confining the alteration within an area that will become the orebody.

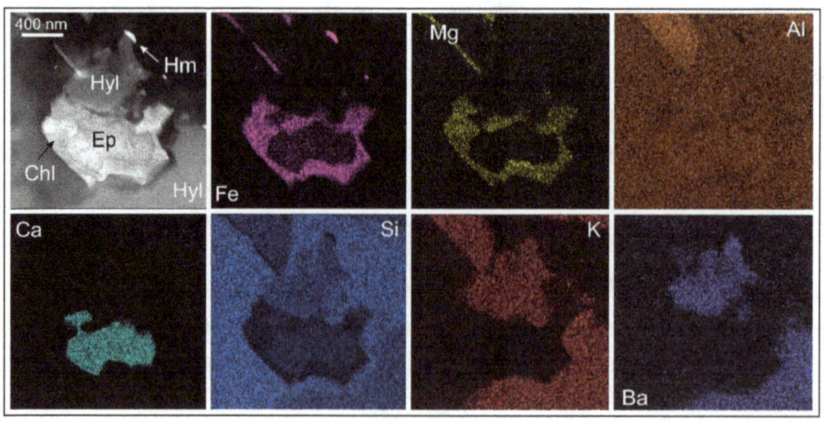

Given figure indicates that HAADF STEM image (top left) and energy-dispersive X-ray-STEM element maps (Fe, Mg, Al, Ca, Si, K, Ba) of ~1_m-sized inclusion of epidote (Ep) replaced by chlorite (Chl) within hyalophane (Hyl). Note the presence of sub-m size hematite (Hm).

In Figure, we show HAADF STEM images and element maps of REE-fluorocarbonates from a locality adjacent to the Olympic Dam deposit. These are part of an assemblage containing sericite and molybdenite pseudomorphically replacing magmatic feldspar (plagioclase) in red-stained K-feldspar altered granite. Bastnäsite and parisite are recognised as the main REE-species, and nanometre-sized hematite is seen as platelets throughout the parisite. Atomic-scale HAADF STEM imaging was undertaken for the first time on such assemblages, and has shown the presence of a third species, "B2S", with composition between bastnäsite and parisite, as well as the changes between the minerals along directions that are perpendicular to the stacking sequences. The element maps in Figure show the changes in parisite-bastnäsite stacking sequences across a microfracture adjacent to areas where hematite is also present as epitaxial intergrowths with minerals of the bastnäsite-synchysite group. This type of assemblage is important for understanding development of the early stages of alteration associated with IOCG deposits, and particularly the link between REE-enrichment and the initial stages of Fe-metasomatism. As in the other example above, this reinforces the hypothesis that the earliest hydrothermal Fe-oxide formed at Olympic Dam is not necessarily magnetite, but hematite. This contradicts generic schemes of IOCG alteration invoking a so-called IOA-stage or "magnetite-apatite" assemblage, which is contended as typifying "early mineralisation" at Olympic Dam, or indeed in IOCG and related "IOA" systems elsewhere.

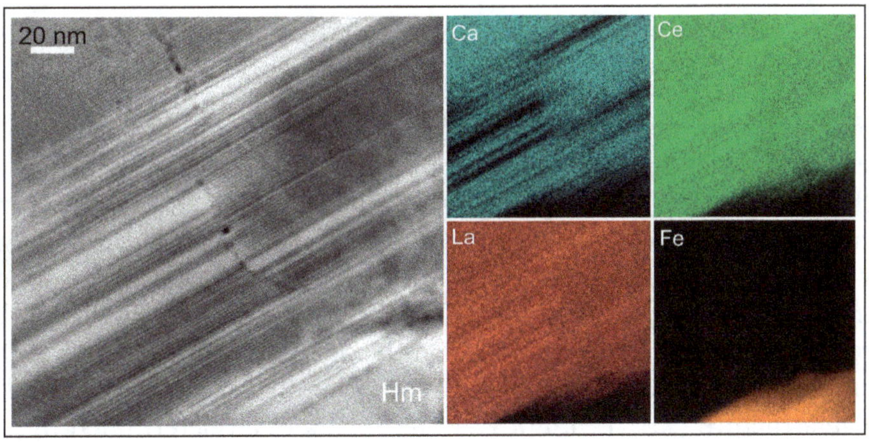

Given figure indicates that HAADF STEM image and STEM-EDS element maps (Ca, Ce, La, Fe) showing lattice-scale intergrowths between bastnäsite and parisite. Note epitaxial hematite is also present.

Electron Backscatter Diffraction (EBSD)

Since the technique first emerged in the mid-1990, EBSD has found wide application in mineralogy, with particular suitability to more refractory phases, such as zircon and garnet. Among the ore minerals, pyrite has been a major focus. Many studies used EBSD to target mechanisms of sulphide deformation at different conditions in metamorphic environments. Other studies used EBSD to address the growth of colloform and framboidal pyrite. Reddy and Hough investigated Witwatersrand pyrite, linking trace element distribution patterns to sulphide deformation, and showed that some of the genetic controversies on the Witwatersrand gold ores can be explained with a model of gold mobilization before and after the re-deposition of pyrite. Iron-(Ti)-oxides are also refractory minerals in which genetic and deformational history can be preserved. Several EBSD studies have been undertaken in rocks ranging from gabbros and deformed granites to marine sediments. Barbosa and Lagoeiro, used EBSD to address transformation between magnetite and hematite in samples from the Quadrilátero Ferrífero of Brazil, while Rosière et al. assessed the interplay between deformation and mineralization in high-grade iron ores hosted by shear zones. Based on synthetic deformation experiments showing the crystallographic textures may be weak or slow to develop, caution has been raised regarding the interpretation of magnetite deformation mechanisms. In recent years, however, researchers have begun to use EBSD to investigate a range of other ore minerals, such as pyrrhotite-pentlandite from komatiite-hosted ores, in which trace element concentrations were found to correlate with microstructures, indicating intragrain diffusion.

EBSD has also proven most helpful in examining potential variation in isotopic signature with crystal orientation of iron-oxides, rutile, or cassiterite. Because electron backscatter patterns are generated from the uppermost few tens of nanometres of the sample surface, preparation and chemical polishing routines need to be optimised. In our experience, this can be problematic for some ore minerals or intergrowths thereof where conventional grinding and polishing causes damage to crystal lattices. Increasingly, EBSD is run on a FIB-SEM platform, offering high-resolution imaging and the potential of three-dimensional tomography. FIB-SEM based EBSD also offers the possibility of working on ion-milled surfaces, negating the need for chemical polishing. Investigating single crystals of uraninite by EBSD on a FIB platform, Macmillan et al. addressed how lattice strain could be correlated with changes in chemical zonation patterns.

Trace Element X-ray Fluorescence (SXRF) Mapping using Synchrotron Radiation

The resolving power of synchrotron radiation and increased availability of access to synchrotron facilities has generated new avenues for investigation of ore minerals. Megapixel high-definition trace element X-ray fluorescence (SXRF) mapping using the Maia detector, for example, enables the generation of maps with the dimensions of a full thin section or drillcore fragment at resolutions of a few μm in only a few hours, complementing EPMA or LA-ICP-MS analysis. Applications of SXRF mapping in ore mineralogy and petrology are given by. We would also single out the recent use of synchrotron X-ray sources for the determination of the oxidation states of minor elements (~1000 ppm) in common sulphides via extended X-ray absorption fine structure (EXAFS) and X-ray absorption near edge structure (XANES). For example, Cook et al. determined the presence of Cu^{+} in indium-bearing sphalerite, confirming $2Zn^{2+} \leftrightarrow Cu^{+} + In^{3+}$ coupled substitution. A second study confirmed the presence of Ge^{4+} in sphalerite, which is suggested to substitute for $(Zn,Fe)^{2+}$ with vacancies introduced to achieve charge balance. Most recently, Bonnet et al. Presented evidence for both Ge^{2+} and Ge^{4+} in sphalerite. Opportunities exist to use such synchrotron source methods to validate the oxidation states of other proposed substituents, e.g., Sn in common sulphides.

Automated Mineralogy

Automated mineralogy techniques on a SEM platform, notably Quantitative Evaluation of Mineralogy via Scanning Electron Microscopy (QEMSCAN) and Mineral Liberation Analysis (MLA), have proven invaluable for the investigation of ore samples and references therein. QEMSCAN uses signals from the BSE and EDS detectors to identify each particle based on average atomic mass, and a comparison of the obtained image with a “mineral library” database to search for probable fits for the compound in question. MLA uses grey-scale BSE images as a primary method of mineral identification and is faster, since the dead time that is required for X-ray analysis is minimized. These methods allow for the determination of which minerals are present, and in which proportions, their grain size distributions and associations. MLA data is especially useful where sample representivity is paramount.

Automated mineralogy has thus found widespread application in quantitative geometallurgy (e.g.,) and routine MLA analysis on large numbers of samples is an essential component of validated deposit-scale zonation models, e.g., for Cu-Fe-sulphide species at Olympic Dam. Automated mineralogy is increasingly used for the identification of unconventional or less common ore minerals and to understand the distributions of minor and trace elements. For example, describe new strategies, combining MLA and EPMA, for the systematic detection and identification of small particles containing the critical element indium. Correct interpretation of automated mineralogy data does, however, require a sound mineralogical understanding of the material, and thus should not be considered as a direct substitute for that. MLA may also not necessarily detect phases present as smallest phases, or resolve subtle compositional variations.

Nanoscale Secondary Ion Mass Spectrometry (NanoSIMS)

NanoSIMS offers in-situ visualisation of mass abundances in solid materials with chemical imaging down to the nanoscale. There are currently ~40 nanoSIMS instruments worldwide, and the technique is proving extraordinarily valuable for understanding the internal structure and composition of ore minerals. Among ore minerals, pyrite and arsenopyrite have been a major focus in published work, particularly with respect to their capacity for hosting both nanoparticulate and solid solution gold. Opportunity exists to use nanoSIMS to address other outstanding issues in the mineralogy of compositionally-zoned minerals, trace element partitioning, and in mineral processing. In one innovative application of nanoSIMS, Wacey et al. have used nanoSIMS mapping of organic carbon and nitrogen in framboidal pyrite, elucidating mechanisms of framboid formation in terms of organic biofilms.

NanoSIMS is proving a frontier technique for tracking the grain-scale distributions of daughter products of uranium decay in ores. This approach has shown both the versatility and extreme sensitivity of the method, even at extremely low concentrations, and points to the future promise of nanoSIMS to constrain element and isotope mobility at the grain-scale, with major implications for improved microbeam geochronology. NanoSIMS has been used to map trace element distributions in minerals that are zoned at the nanoscale, including datable minerals such as zircon or monazite.

Atom Probe Tomography (APT)

APT has, until recently, been little applied to ore minerals, despite being around for more than a decade and widely used by material scientists and others to map the spatial distribution of atomic species in three dimensions and at the nanoscale. APT couples chemical sensitivity and three-dimensionalm capabilities with a spatial resolution near the atomic scale to provide unique information about nanoscale structures. The technique has been demonstrated as a valuable tool to complement SIMS and allow for smallest-scale geochronology to understand nanoscale diffusionwithin zircon, and for the examination of homogeneity within geochronological reference materials.

The potential that this unique method may have in mineralogy. Weber et al. used APT to demonstrate a hitherto-unrecognised layering of nanoscale pore structures in barites, which carries implications for the observed reactivity of barite and rapid uptake of Ra. In another APT study, Parman et al. address natural isoferroplatinum (Pt_3Fe), showing potential order-disorder distribution of trace elements. The considerable potential of APT as a breakthrough method that may help to resolve outstanding issues that connects structural features as observed by TEM with geochemical features. Introducing APT to a mineralogical audience, Parman notes that the method "forces one to merge mineralogy and geochemistry", and that APT "should stimulate a range of new discoveries".

Geochronology using Ore Minerals

The accurate determination of the timing of mineralization has been a long-standing issue in mineral deposit studies. Effective dating is frequently hampered by a mismatch between those minerals that can be accurately and reliably dated, and those which are abundant in ores and can be petrographically constrained as part of, and therefore contemporaneous with mineralization. Molybdenite Re-Os geochronology has proven an invaluable tool to navigate this impasse, and has been widely applied to the many cases where molybdenite is present within the ore assemblage. More recently, other non-traditional minerals have been shown to have utility in geochronology, notably hematite, which, in some ore systems, readily incorporates U and radiogenic Pb into its crystal structure. These studies have shown that hematite can be accurately and precisely dated by LA-ICP-MS, sensitive high-resolution ion microprobe (SHRIMP), or isotope dilution thermal ionization mass spectrometry (ID-TIMS). Moreover, hematite and magnetite appears remarkably robust and do not undergo extensive metamictization. U-Pb dating of iron-oxides by LA-ICP-MS can provide rapid, reliable results, at a sufficient resolution to be of value in reconnaissance studies.

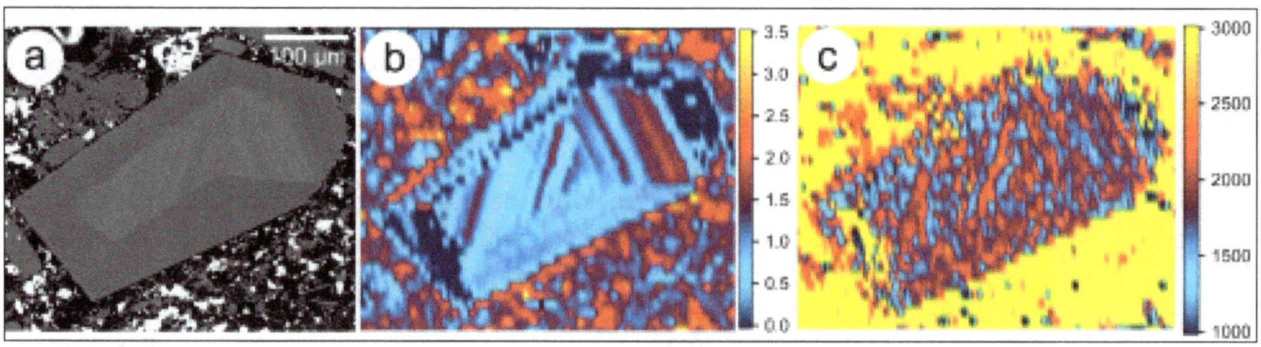

Figure shows (a) Indicates that BSE image of a U-bearing, compositionally-zoned hematite grain (Olympic Dam); (b) LA-ICP-MS concentration map for U (scale in counts-per-second, logarithmic scale, 10n); and (c) Map showing 206Pb/238Pb ratio and scale in in Ma. The yellow background represents high 204Pb in the surrounding matrix mineralogy.

Geochronological data can also be obtained on several other ore minerals, including columbite-tantalite, rutile, thorite and uraninite by LA-ICP-MS. LA-ICP-MS "age" mapping of radiogenic isotopes by rastering across a selected area of a grain using appropriate reference standards in zircon, or hematite as demonstrated in Figure. Such methods also have potential for the assessment of homogeneity in minerals with zonation patters relative to U and radiogenic Pb. The ability to investigate minerals and visualise element and isotope distributions at the nanoscale, at the nanoscale, using LA-ICP-MS mapping, nanoSIMS, APT, and other techniques, with complementary structural information obtained via EBSD, have highlighted the heterogeneity present in natural samples and the possibility of intra-grain diffusion, as well as the need to ensurethat reference standards are homogeneous as possible.

Non-Traditional Stable Isotopes

The growing interest in in-situ analysis of "traditional" stable isotopes notwithstanding development and wider availability of laser-ablation multi-collector (LA-MC)-ICP-MS instrumentation has catalysed a rapid expansion in analysis of non-traditional stable isotopes, including the isotopes of several elements concentrated within ores and measured in ore minerals. These include iron, nickel, copper, zinc, germanium, selenium, and molybdenum. Several other isotopic systems with potential application to understanding the evolution of ore deposits are rapidly emerging. These include thallium, tellurium, and mercury.

Table: Applications of stable isotope geochemistry in ore mineralogy.

Isotope System	Key Ore-Related Questions Addressed
Fe	Origin of BIFs; metal-silicate, sulphide-melt, fluid-mineral and various mineral-mineral fractionations; 56Fe as a tracer of ore-forming processes, including seafloor hydrothermal vents; discrimination of hydrothermal vs. magmatic processes; origin of detrital pyrite in Archaean sedimentary rocks.
Ni	The origin of komatiite- and laterite-hosted nickel ores.
Cu	Redox-induced fractionations wide large variability in supergene ores; large potential for application to complex deposits, e.g., IOCGs.
Zn	Less data than for Cu-isotopes. Evidence for modest fractionation during evolution of hydrothermal fluids.
Ge	Ge concentrated in sphalerite; potential for tracing processes in seafloor hydrothermal systems.
Se	Selenium source; distinguishing magmatic and hydrothermal input in ancient deposits.
Mo	Significant application to ore-forming processes in deposits containing molybdenite; fractionation induced by redox reactions, fluid boiling.
Te	Ore-forming processes in Te-bearing epithermal and orogenic precious metal deposits.
Hg	Relatively little used in ore geology to date but with potential application to hydrothermal systems.
Tl	Significant isotopic variability in natural samples points to the potential of Tl-isotopes for understanding formation of hydrothermal deposits; more baseline data are needed.

New Approaches and Sub-Disciplines

The recent advances outlined above have, however, not only been brought about by rapid developments in analytical capability, but also by an evolution in conceptual approaches to ore mineralogy.

Trace Elements in Ore Minerals

Very few minerals are ever "pure"; many can incorporate two dozen or more different elements at measurable and quantifiable concentrations—and furthermore, many minerals can be compositionally zoned. Interest in the content and distribution of minor and trace elements in ore minerals has expanded rapidly for several reasons, driven by analytical capabilities that are highlighted above. Firstly, the precise in-situ measurement of trace element concentrations has direct economic implications since it enables the mineralogical distribution of elements that represent either potential economic by-products or harmful elements in an ore potentially decreasing its value, to be accurately pinpointed. This type of knowledge may directly or indirectly impact upon plant design for mineral processing. Secondly, accurate information on the mineral site(s) of potentially process deleterious elements provides data that can assist in management and long-term storage of mining or processing waste. Thirdly, trace element analysis can assist in understanding the genetic history of complex ore deposits. The recognition of distinct generations of a given mineral, or of grain-scale compositional zoning, can, for example, help to track the evolution of hydrothermal fluids, or expose evidence for an overprinting event that has implications for metal distributions within the deposit. Such an understanding can, in turn, underpin metallogenetic models, with direct application to model-based mineral exploration. Lastly, spatial patterns of trace element distribution within specific minerals can be potentially applied in exploration vectoring, offering greater sensitivity than whole-rock analysis.

The Significance of Gangue Minerals—A Holistic Approach to Ore Deposits

The gangue components of an ore mineral assemblage are increasingly recognised as rich sources of information to support models of ore formation and petrogenesis. Some particularly refractory minerals may, in fact, preserve textural and compositional features that reveal deposit evolution better than many ore minerals. In our laboratory, these have ranged from garnets and other calc-silicate minerals in skarns], through fluorapatite and REE-fluorocarbonates, to alkali and plagioclase feldspars. A detailed investigation of hematite, which contributes as much as 80 wt % of the ore at Olympic Dam, has shown how this mineral can encode a wealth of valuable trace element information.

This approach has led to the recognition that to adequately understand the distribution of those elements, which are particularly valuable for petrogenetic interpretation, notably REE, or the high field strength elements, a holistic approach is needed, in which all of the minerals, major and minor, in the system are studied as potential hosts. Partitioning among coexisting minerals can impact significantly on geochemical signatures, and changes in partitioning behaviour may itself provide insights into the evolution of mineral deposits. Holistic LA-ICP-MS strategies have also revealed how elements of interest (potential by-products and process deleterious impurities) may partition among coexisting hydrothermal base metal sulphides, for example, between galena, sphalerite, chalcopyrite, and tetrahedrite-tennantite. This type of research has also shown how some gangue minerals can host unexpectedly high concentrations of elements of economic interest; concentration of U, W, Sn and Mo in skarn garnets representing one good example. Several opportunities currently exist to confirm whether significant resources of so-called critical metals occur within new resource types, e.g., gallium in epidotites. These represent exciting research areas in which grain-scale analysis by LA-ICP-MS is expected to play a key role.

We have shown how establishing models of trace element distribution in mineralogically complex ore systems need to acknowledge not only the major sulphide and oxide ore minerals and rock-forming silicates, but also relatively minor phases that can sequester some trace elements at significant concentrations. In the example from Fe-Ti-V ores from the Panzhihua layered intrusive complex, China, and, as shown as Figure, trace elements are distributed between rock-forming silicates (pyroxene, olivine, feldspar), titanomagnetite (the dominant oxide), and sulphides (pyrrhotite and minor chalcopyrite and cobaltpentlandite). Exsolution of ilmenite and spinel from titanomagnetite has, however, sequestered certain trace elements. Importantly, the high field strength elements (Sc, Zr, Hf, Ta, and Nb) are all markedly enriched in ilmenite relative to titanomagnetite, whereas Zn, Ga, and others, occur preferentially in spinel exsolutions within titanomagnetite.

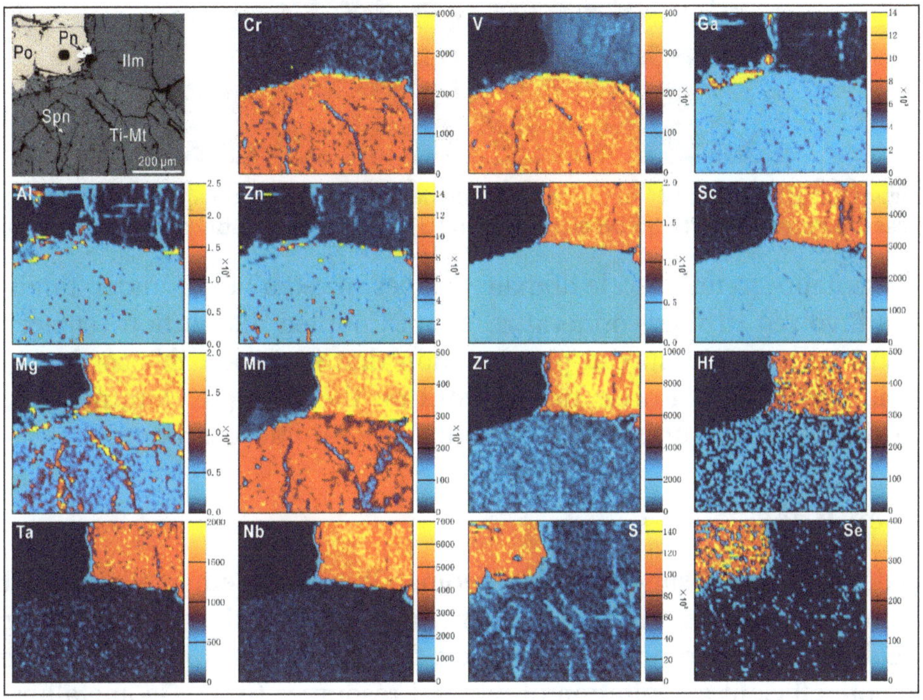

This figure indicates LA-ICP-MS maps of assemblages in Fe-Ti ore from the Panzhihua intrusion, southeast China: Ti-Mt: titanomagnetite, Ilm-ilmenite, Spn-spinel, Po-pyrrhotite, Pn-pentlandite. Note the relative enrichment of Ti, Sc, Mg, Mn, Zr, Hf, Ta, and Nb in ilmenite relative to titanomagnetite.

Solid Solutions and Nanoparticles

The capability to explore smaller and smaller volumes of sample material has opened up exciting opportunities to understand the nature and behaviour of ore minerals at the nanoscale. Visualisation of the nanoscale features of minerals allows for the verification of crystal structures, polytypism and/or polysomatism in mixed-layer compounds, etc. Moreover, atomic-scale microscopy can help address fundamental questions on the nature of sold solution and how minerals adapt to incorporation of trace and minor elements. Nanoscale studies have also confirmed the role played by mineral nanoparticles, the stability and reaction behaviour of which are governed by very different sets of thermodynamic rules compared to their larger counterparts.

The example in Figure shows hematite from the Olympic Dam orebody. This contains not only nanoparticles of uraninite (U-NPs), but also micron-scale oriented lamellae of REE-fluorocarbonates (bastnäsite-parisite group), which are interpreted to have coarsened up during recrystallization of granitophile (U-W-Sn-Mo-bearing)-rich hematite interacting with fluids via CDR reactions. Such reactions evolve via transient porosity providing sites for mineral deposition, as in this case U-NPs. The source of U is from hematite, which can incorporate up to hundreds of ppm of these "granitophile" elements.

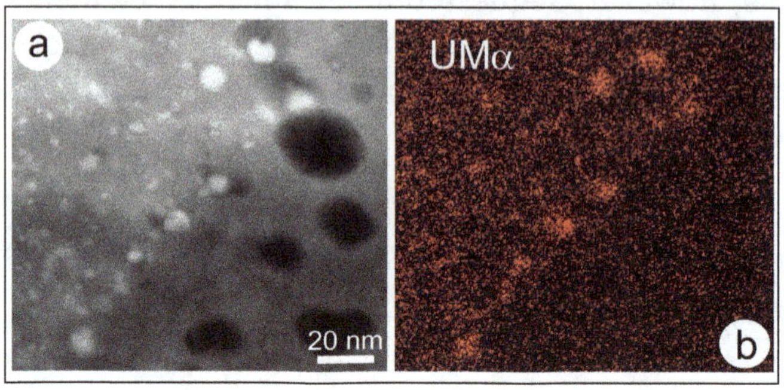

Figure: (a) HAADF-STEM image of uraninite nanoparticles (bright spots) in hematite. Dark areas are Pores; (b) STEM-EDS element map (U) confirming identity of the bright inclusions.

Accessibility to the nano- to microscale domain has permitted a better understanding of the critical mechanisms involved in fluid–mineral reaction. Coupled dissolution replacement reaction (CDRR) is recognised as a widespread phenomenon that is responsible for local-scale mineral transformation in hydrothermal ores. In our work on the Olympic Dam deposit, evidence for CDRR is recorded in feldspars, uraninite, hematite and REE-fluorocarbonates.

New Sub-Disciplines

In the past two decades, a new sub-discipline, geometallurgy, has emerged, in which geology and mineralogy is combined with extractive metallurgy and geostatistics. Geometallurgy enables the derivation of geologically-based predictive models for minerals processing that can assist plant design and promote efficient exploitation of mineral resources (e.g.). Geometallurgy requires an interdisciplinary approach and widespread application of automated mineralogy, expanding traditional ore mineralogy into new spheres of economic relevance. Geometallurgical approaches are increasingly applied to a wide range of different deposit types.

Discovery of New Minerals

Despite only making up a tiny fraction of the Earth's crust, mineral deposits have proven to be rich sources for the discovery of new mineral species. In fact, if we include samples from abandoned mines and oxidised ores at the surface associated with historical mine workings, mineral deposits account for roughly half of the new minerals that were accepted between 2011 and 2016. This is perhaps unsurprising as mineral deposits are anomalous in elements that are prone to greater mineralogical diversity and even more importantly, will contain various combinations of those elements.

Sulphosalts represent a particularly fertile area of ore mineral research, with more than a dozen new sulphosalt species being approved since the publication of the report of the IMA Commission on Ore Mineralogy Sub-Committee on Sulphosalts. It is also reasonable to expect that many of the predicted >1563 yet-to-be-discovered mineral species will be identified in samples from existing or historical mine sites. Localities such as Tsumeb (Namibia), Långban (Sweden), Lengenbach (Switzerland), Shinkolobwe (Democratic Republic of Congo), and Franklin (NJ, USA) all feature in a listing of the top 10 localities in terms of the number of minerals, for which that locality is registered as the type locality, with 72, 72, 43, 38, and 35 mineral species, respectively. The technological revolution outlined above has accelerated the search for, and characterization of new ore minerals, particularly in cases where the phase in question is only a few microns in size and intergrown with other related phases. For example, FIB-SEM was successfully used to extract tiny samples of the new mineral gra¸tianite for structural study.

Hazen et al. have recently made a convincing case for the acknowledgement, as "anthropogenic minerals", of compounds that originate from human action. The majority of the 208 species put forward as anthropogenic mineral candidates can be attributed to mining-related activities, and are found within ore dumps, in slags and their weathering products, in mine shafts and adits, and on smelter walls.

Challenges

Ore mineralogy and ore microscopy remain highly relevant. Underteaching of traditional ore microscopy techniques in the World's higher education institutions is perhaps a reflection of the relative ease of access to advanced microanalytical instrumentation, which offer the advantage of compositional information. The need for innovative, interdisciplinary approaches in the search for mineral resources to guarantee future security of supply means that not only geologists, but also metallurgists and processing engineers, need to appreciate the pivotal role that is played by ore mineralogy in ensuring the sustainable exploitation of limited resources. Mineral characterization remains a critical part of any study, so that reliable interpretations can be made based on knowledge of exactly what has been analysed.

Descriptions of ore minerals, new occurrences, and information on mineral textures or paragenesis, remain important, as do compilations of compositional or other types of quantitative data. Ores are by essence ephemeral in character, and once mined, are gone forever. Museums and other national, state, or municipal institutions have a vital role to play in archiving particularly interesting or instructive sample material. Likewise, universities need to ensure that teaching and research collections are maintained to provide students with representative material from typical "run of mine" ore. Reference collections should be assembled in collaboration with the mining companies at the time of mining, and maintained as a resource for future research.

Identifying current frontiers in ore mineralogy is a subjective exercise. Obtaining quantitative compositional data on the smallest particles is one obvious frontier that can be crossed with contemporary instrumentation. The increased focus on smaller and smaller scales of observation, and not infrequently on smaller sample suites due to restricted access, or the high cost and time-consuming nature of the more advanced microanalytical techniques, does, however, generate unease about just how representative the investigated samples are. This may, in turn, impact on the reliability of broad interpretations that are based on empirical observation on only the tiniest fractions of an orebody. Further in the future, the widespread use of automated mineralogical techniques and large microanalytical datasets will likely trigger a new revolution in which machine learning can be expected to play an increasing role.

Many of the techniques and approaches discussed here complement one another and can often be used in combination to achieve in-situ characterisation of the same sample at different observational scales and references therein. Most methods do, however, have their limitations, in terms of sensitivity, resolution, or less quantifiable factors, such as ill-constrained sources of error, a lack of suitable standard materials, and so on.

Chapter 2

Mineral Chemistry

Minerals are naturally occurring solids with a definite but not necessarily fixed chemical composition. Years of weathering and erosion exposes them to the elements of nature which interact and modify their chemical compositions. This chapter deals with mineral chemistry and its related aspects like metamorphism and metasomatism in detail.

Minerals can form by any of the following processes:

- Precipitation from a fluid like H_2O or CO_2: This can take place within the Earth by hydrothermal processes, diagenesis, and metamorphism, and at or near the Earth's surface as a result of evaporation, weathering, or biological activity.
- Sublimation from a vapor: This process is somewhat rarer, but can take place at a volcanic vent, or deep in space where the pressure is near vacuum.
- Crystallization from a liquid: This takes place during crystallization of molten rock (magma) either below or at the Earth's surface.
- Solid - Solid reactions: This process involves minerals reacting with other minerals in the solid state to produce one or more new minerals. Such processes take place during metamorphism and diagenesis due to changing temperature and pressure conditions.

No matter what process is involved, a particular mineral cannot form unless the chemical ingredients necessary to make the mineral are present. Thus, the most common minerals are minerals that have a chemical composition made of the common elements found in their environment. Since the environment where most observable minerals occur is the Earth's crust, we must first explore the chemical composition of the Earth's crust.

Composition of the Crust

The Earth's crust is made up of about 95% igneous and metamorphic rocks, 4% shale, 0.75% sandstone, and 0.25% limestone. The continental crust has an average composition that approximates granodiorite (a medium to siliceous igneous rock), whereas the oceanic crust has an average composition that is basaltic (a low silica igneous rock). Based on information, the average composition of the continental crust is as follows:

Element	Wt%	Atomic%	Volume%
O	46.60	62.55	~94
Si	27.72	21.22	~6
Al	8.13	6.47	
Fe	5.00	1.92	

Ca	3.63	1.94	
Na	2.83	2.34	
K	2.59	1.42	
Mg	2.09	1.84	
Total	98.59	100.00	100

8 elements make up over 98% of the Earth's crust and that Oxygen is the most abundant element. This becomes even more evident if the elements are determined on an atomic basis, where we can see that about 63 out of every 100 atoms in the crust are Oxygen. On a volume basis, Oxygen makes up about 94% of the crust because Oxygen is a large anion,m and the other elements occur as small cations coordinated by the Oxygen anions.

Because of the average composition of the crust, the most common minerals found in the crust are silicates and oxides. Of the silicates, the aluminosilicates, like the feldspars and clay minerals are the most common.

Other minerals, containing the other elements in the periodic table are found in the crust. But, in order for these minerals to occur, special geologic conditions are necessary to concentrate these less abundant elements so that they occur in high enough concentrations to form a separate mineral. Elements that occur in concentrations less than about 0.1% (usually measured in parts per million or parts per billion) are called dispersed elements or trace elements.

Methods of Chemical Analysis

While the common minerals are made up mostly of the abundant elements in the Crust, and we can express the chemical composition of such minerals by a simple (and sometimes not so simple) chemical formula, all minerals exhibit some chemical variation. In fact pure minerals (those with a chemical composition exactly the same as the chemical formula) are rarely found in nature. We will discuss why this is the case later on in the lecture. For now, it is sufficient to say that if we had a technique for measuring even the smallest concentrations of elements, we would see that any given mineral probably contains every stable element in the periodic table. Of course most of these elements would occur in very low concentrations (parts per trillion or less), nevertheless they are present.

Various methods have been used to perform chemical analyses of rocks and minerals. But, analysis of minerals is quite different from analysis of rocks. Chemical analysis of rocks involves analyzing an aggregate of minerals. So, in most case the rock can be ground to a fine. Analysis of a mineral is more complicated because:

- Individual minerals are much smaller than rocks and it is therefore difficult to obtain enough sample of a mineral to perform the chemical analysis.
- In cases where a single mineral is too small to provide enough sample, several grains of the same mineral can be separated from a rock and analyzed as a single sample. But, this involves extensive work in separating the mineral form all other minerals, and still does not tell us if there is any chemical variability between individual grains. (The electron

microprobe technique is an exception, because it allows for analysis of small areas of single grains).

- Individual minerals may be chemically zoned: That is there may be differences in the chemical composition of the mineral from its center to its rim. Chemical analysis of single mineral grains or aggregates of mineral grains will not reveal this chemical zonation. (Again, the electron microprobe technique is an exception).

Wet Chemical Analyses

Prior to about 1960 classical wet chemical analysis was the only means of determining the chemical composition of minerals. Such analyses were limited by the factors in terms of sampling size, although techniques were developed for analyzing very small samples. Wet chemical analyses always involve dissolving the mineral into an acidic solution. In order for the dissolution to take place completely, the mineral is usually first ground into a fine powder (to increase surface area) and the appropriate acid must be used. Wet chemical analysis can be classified into three different types:

- Gravimetric Analyses: In this case the element of interest is precipitated as a compound. The precipitate is then weighed to determine its proportion in the original sample.
- Volumetric Analyses: In this type of analysis titration is used to determine the amount of reagent that is added in order for a specific chemical reaction to occur that involves the element of interest. From the volume of reagent added, the concentration of the element can be calculated.
- Colorometric Analyses: A reagent is added to the solution that reacts with the element of interest to produce a color change in the solution. The intensity of the color is proportional to the concentration of the element of interest, and thus when compared to standard solutions in which the concentration is known, the concentration of the element in the unknown solution can be determined.

Generally all elements cannot be determined by the same wet chemical methods and different methods are more sensitive for different elements. Thus, complete wet chemical analysis usually involves a combination of methods. Wet chemical analyses require a skilled analytical chemist, and are usually very time consuming.

Inductively Coupled Plasma - Mass Spectrometry (ICP-MS)

ICP-MS combines two instruments, an inductively coupled plasma spectrometer and a mass spectrometer to simultaneously measure most elements in the periodic table and determine concentration down to the sub part-per trillion (ppt) levels. The sample of mineral or rock must first be powdered and put into solution in acid. This solution is then injected into plasma as an aerosol where the atoms in the sample are ionized. The plasma is produced by passing Ar gas into an energy field induced by a radiofrequency (RF) generating coil. The RF energy field causes collisions between the Ar atoms, generating a high energy plasma. The sample aerosol is instantly decomposed in the plasma (temperature of 6,000 - 10,000 K) to form analyte atoms which are simultaneously ionized. The ions produced are extracted from the plasma into a mass spectrometer which is held at high vacuum. The mass analyzer then separates the ions based on their mass/charge ratio. The mass analyzer is essentially four parallel molybdenum rods to which a combination of RF and DC voltages are applied. The combination of these voltages allows the analyzer to transmit only ions of a specific mass/charge ratio. Finally, the ions are measured using an electron multiplier, and are collected by a counter for each mass number. The mass spectrum generated is extremely simple. Each elemental isotope appears at a different mass (i.e. 27Al would appear at 27 amu) with a peak intensity directly proportional to the initial concentration of that isotope. The proportionality is determined using standard solutions with known concentrations of elements.

Small parts of solid mineral samples can also be analyzed using the ICP-MS with an attached laser. The laser is focused on the part of the sample of interest and the power is increased to cause melting of that small part of the sample (this is called laser ablation). The melt is then injected into the plasma and the rest of the analysis is similar to that described above. This is a relatively new technique that is continually improving in terms of detection limit and reproducibility.

X-ray Fluorescence (XRF) Spectrometry

XRF spectrometry uses some of the same principles we discussed in our lecture on X-ray diffraction. It is a technique now used mainly for the analysis of rocks because it requires relatively large amounts of sample (about 1 gram). It can be used to measure all of the major elements in a rock or mineral as well as a wide variety of trace elements. Sample preparation involves grinding the rock or mineral to find powder and either pressing the powder into a pill or melting and quenching the sample to make a glass disk.

When an electron is be dislodged from an inner electron shell, outer shell electrons fall in to take its place resulting in the production of X-rays with a wavelength that is characteristic of the element involved. In X-ray diffraction, a beam of electrons is used to dislodge the inner shell electrons and to produce X-rays of known wavelength in the anode target of the X-ray tube. In XRF, however, we use the X-rays from the X-ray tube to dislodge the electrons in atoms of an unknown sample.

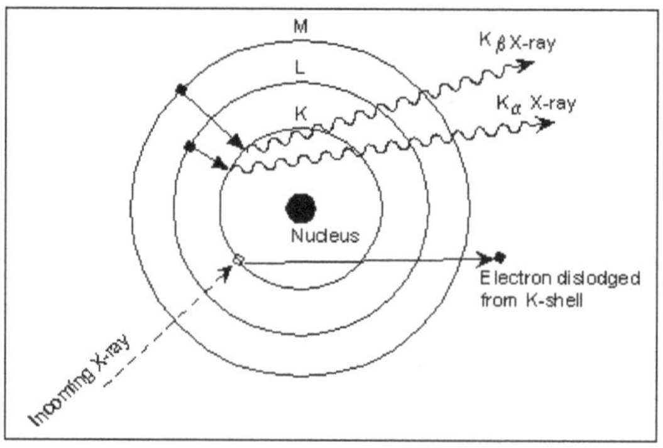

Just like in the X-ray tube, inner shell electrons that are dislodged from atoms in the unknown sample result in the production of X-rays of characteristic wavelength for each of the different atoms in the sample. For each atom in the sample, the intensity of the characteristic X-rays will be proportional to the number of atoms in the sample. Thus, if we can measure the intensity of characteristic X-rays for each element, we can compare the intensity to that of a standard with known concentration, and determine the concentration of the element in the unknown.

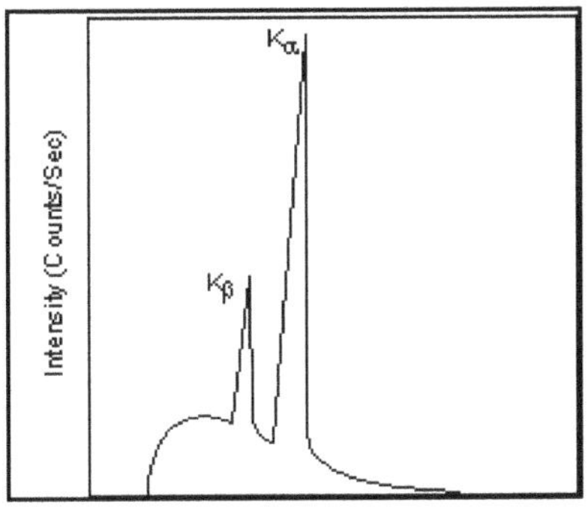

The instrument used for such measurements is called an X-ray Fluorescence Spectrometer. It consists of an X-ray tube to produce X-rays capable of dislodging inner shell electrons of all elements of interest in the sample These have to be high energy X-rays, so Cr, Mo, W, and Au targets are used as the anode. The X-rays that are produced in the sample are then sent through a collimator to produce a coherent beam. These are then diffracted through a crystal of known d-spacing. Using the Bragg equation ($n\lambda = 2d\sin\theta$) we can determine the Tangle that Xrays of known wavelength will be diffracted through the crystal. Thus, the crystal and detector are set up to diffract only X-rays of the particular wavelength of interest into the detector.

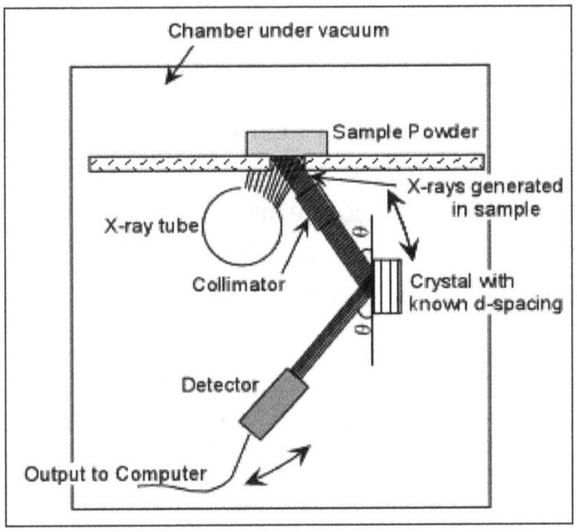

Intensities of characteristic X-rays for each element of interest are determined by rotating the crystal and detector to the angle θ required for diffraction of the characteristic X-ray wavelength. X-ray intensities are then measured for each element in the unknown and for each element in a standard that has known concentrations. The following equations are then used to determine the concentration of each element in the unknown:

$$Ci^{std} = kI_i^{std}$$

$$Ci^{unk} = kI_i^{unk}$$

where C_i^{std} = The concentration of element i in the standard,

I_i^{std} = The intensity of X-rays of element i in the standard,

C_i^{unk} = The concentration of element i in the unknown,

I_i^{unk} = The intensity of X-rays of element i in the unknown, and

k = A proportionality constant.

The proportionality constant, k, can be determined from the first equation:

$$k = C_i^{std}/I_i^{std}$$

This can then be substituted into the second equation to solve for the concentration of element i in the unknown:

$$C_i^{unk} = (C_i^{std}/I_i^{std})\ I_i^{unk}$$

In practice, this is only true if all elements in the standard and the unknown occur in about the same concentration. This is often difficult to attain, so a wide variety of standards are used to figure out all of the interfering effects and make a calibration curve which relates concentration in the unknown to a more complex equation involving the concentration and intensities in the standard.

Electron Microprobe (EMPA) Analysis

The electron microprobe is an instrument that uses principles similar to XRF. The main difference is that instead of using X-rays to dislodge inner shell electrons of elements in the unknown, an electron beam is used. The electron beam can be focused down to a spot size of about 1 Pm (1Pm = 10^{-6}m), and thus the X-rays in the sample will only be generated within a very small volume of the sample. Thus, one can measure the concentrations in small parts of single mineral grains, and chemical zoning patterns or changes in the concentration of elements through an individual grain can be determined. The sample can either be a highly polished thin section of a rock or mineral or can be grain mounts wherein individual grains are mounted in epoxy and polished to a smooth surface.

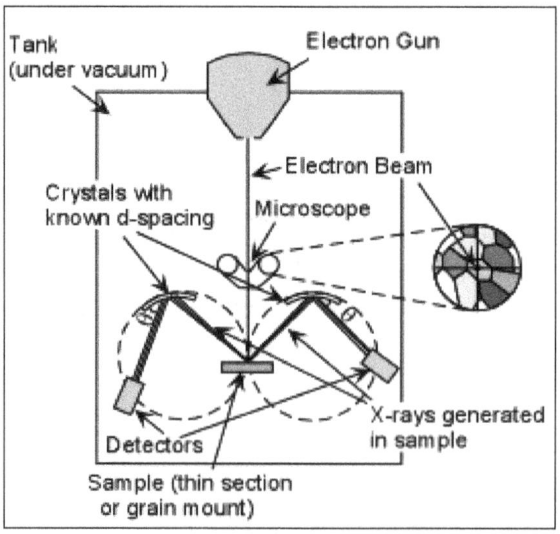

The instrument consists of a large tank that is evacuated to produce a vacuum. At the top is an electron gun that has a filament, similar to that in an X-ray tube that generates electrons.

The electron beam is accelerated down to a sample stage that is moveable. The electron beam is focused to a diameter of about 1 Pm by a series of magnets along its path. Most electron microprobes have at least 3 X-ray spectrometers inside, so that more than one element can be measured at once. A microscope with binocular viewfinders allows the user to focus the electron beam anywhere on the sample, as shown in the inset. This can be accomplished by moving the sample stage.

Electrons striking the small area of the sample produce X-rays with wavelengths characteristic of the elements in the sample. The spectrometers consist of curved crystals of known dspacing. These crystals diffract X-rays of each element into the detectors at the diffraction angle T. Several elements can be measured at once, and then the spectrometers are set to new angles to measure the X-ray intensities of several more elements.

Just like in XRF analyses, standards have to measured to determine the proportionality constants or calibration curves, and equations similar to those used in the XRF method are used to determine element concentrations in the unknowns. Electron microprobe analysis is useful for determining the major element concentration in small parts of minerals. But, it is much less sensitive than the XRF in terms of trace element concentrations.

Compositional Variation in Minerals

In our definition of a mineral we said that a mineral has a definite, but not necessarily fixed chemical composition. Here we explore the "not necessarily fixed" portion of the definition. Chemical compositional variation in minerals is referred to solid solution. Although most of us think of solutions as a liquid containing dissolved ions, solids can form solutions as well, in which case we think of one solid as being dissolved in another solid.

Solid solution occurs as the result of ions substituting for one another in a crystal structure. The factors that control the amount of solid solution that can take place in any given crystal structures are:

- The size of the ions and the size of the crystallographic sites into which they substitute. Generally ions of about the same size can substitute for one another, although the size of the crystallographic site can also play a role if one of the ions is of nearly the same size, but is too large to fit into the site.
- The charges on the ions that are substituting for one another. If the charges are the same, then the crystal structure can remain electrically neutral. If the charges are not the same then other substitutions must take place to maintain charge balance.
- The temperature and pressure at which the substitution takes place. In general there is a greater amount of substitution that takes place at higher temperature. This is because the atoms vibrate at a higher rate and the sizes of crystallographic sites are larger. Pressure can also have an effect because it can change the size of both the crystallographic sites and the ions, thus resulting in different substitutions than might take place at lower pressure.

Three different types of solid solution are recognized - substitutional, interstitial, and omission.

Substitutional Solid Solution

Simple substitution

When ions of equal charge and nearly equal size substitute for one another, the solid solution is said to be simple. Generally if the sizes of the ions are nearly the same, the solid solution can occur over the complete range of possible compositions and the solid solution series is said to be complete. If the sizes are similar, but still very different the substitution may only occur over a limited range of compositions and the solid solution series is said to be partial or limited. Partial or limited solid solution can also occur because the substituting ion does not occur in high enough concentrations in the environment in which the mineral is formed. Some common examples are:

Substitution	Ionic Radii (C.N.) Å		Type	Examples
Fe^{+2} <=> Mg^{+2}	Fe^{+2}(6) 0.78	Mg^{+2}(6) 0.72	Complete High T favors Mg	Olivines: Mg_2SiO_4 - Fe_2SiO_4 Pyroxenes: $MgSiO_3$ - $FeSiO_3$ $CaMgSi_2O_6$ - $CaFeSi_2O_6$ Biotite: $KMg_3AlSi_3O_{10}(OH)_2$ - $KFe_3AlSi_3O_{10}(OH)_2$ Tremolite $Ca_2Mg_5Si_8O_{22}(OH)_2$ - Ferroactinolite $Ca_2Fe_5Si_8O_{22}(OH)_2$
Fe^{+2} <=> Mn^{+2}	Fe^{+2}(6) 0.78	Mn^{+2}(6) 0.83	Complete, but limited by amt. of Mn available	Siderite $Fe(CO)_3$ - Rhodochrosite - $Mn(CO)_3$
Mg^{+2} <=> Mn^{+2}	Mg^{+2}(6) 0.72	Mn^{+2}(6) 0.83	Partial	
Na^{+1} <=> K^{+1}	Na^{+1} (8) 1.18	K^{+1}(8) 1.51	Complete at high T Partial at low T	Alkali Feldspars: $NaAlSi_3O_8$ - $KAlSi_3O_8$
Fe^{+3} <=> Al^{+3}	Fe^{+3} (6) 0 .65	Al^{+3} (6) 0.54	Limited	Alkali Feldspar
Br^{-1} <=> Cl^{-1}	Br-1(6) 1.96	Cl-1(6) 1.81	Complete	KCl - KBr
$(OH)^{-1}$ <=> F^{-1}	$(OH)^{-1}$ (4) 1.38	F^{-1}(4) 1.31	Complete	Biotite: $K(Mg,Fe)_3AlSi_3O_{10}(OH,F)_2$

Coupled Substitution

Coupled substitution occurs if an ion of different charge is substituted. This results in having to make another substitution in order to maintain charge balance. Such coupled substitution is common in the silicate minerals where Al^{+3} substitutes for Si^{+4} in tetrahedral (C.N. = 4) sites. Some examples of common coupled substitutions are given below:

Substitution	Extent	Example(s)
$Na^{+1}Si^{+4}$ <=> $Ca^{+2}Al^{+3}$	Complete	Plagioclase: $NaAlSi_3O_8$ - $CaAl_2Si_2O_8$
$Ca^{+2}Mg^{+2}$ <=> $Na^{+1}Al^{+3}$	Limited	Diopside: $CaMgSi_2O_6$ - Jadeite: $NaAlSi_2O_6$
$Mg^{+2}2Al^{+3}$ <=> $2Fe^{+2}Ti^{+4}$	Extensive	Spinels

Another type of coupled solid solution involves filling a site that is normally vacant in order to achieve charge balance. For example, in the amphibole mineral tremolite - $Ca_2Mg_5Si_8O_{22}(OH)_2$, if Al^{+3} replaces one of the Si^{+4} ions then Na^{+1} can go into a site that is normally vacant to maintain charge balance. The resulting formula would be $NaCa_2Mg_5AlSi_7O_{22}(OH)_2$. This would be called a sodic amphibole.

Interstitial Solid Solution

In some crystal structures there are sites that are not normally occupied by ions. These are considered voids. However, when an ion does occupy one of these voids it is called interstitial solid solution.

Omission Solid Solution

Omission solid solution occurs when an ion of higher charge substitutes for an ion of lower charge. In order to maintain charge balance, two of the lower charged ions will be replaced, but the higher charged ion will occupy only one site, thus the other site will become vacant, or omitted. An example of this type of solid solution is found in the blue variety of microcline, in which a Pb^{+2} ions replaces 2 K^{+1} ions. One of the K sites is replaced by the Pb^{+2} and the other site is left vacant.

Exsolution

The extent of solid solution is sometimes dependent on temperature and pressure since the sizes of ions and the sizes of the crystallographic sites can change with temperature and pressure. Thus, some minerals show complete solid solution under one set of temperature/pressure conditions, and only limited solid solution under different temperature/pressure conditions. When the conditions change to those where limited solid solution is favored, the mineral exsolves or unmixes. But, because the process is taking place in the solid state, exsolution or unmixing cannot easily form two separate phases, because the ions must diffuse through the solid. In fact, what happens is that two separate phases form in discrete domains within a single mineral grain. These domains ar crystallographically oriented, so they appear as lamellae or lines across the mineral grain.

Although we will explore this further in our discussion of phase diagrams, an example is given here. The alkali feldspars (albite - $NaAlSi_3O_8$ - orthoclase - $KAlSi_3O_8$) form a complete solid solution at high temperature. At lower temperature the solid solution becomes progressively more limited, and as a result lamellae of albite-rich feldspar begin to grow in the orthoclaserich alkali feldspar. This produces a texture called perthite, where the lighter colored albiterich feldspar is seen to occur as irregular lines or streaks (the lamellae) within the pink orthoclase-rich feldspar.

In the case of perthite, the exsolution lamellae are often large enough to see with the naked eye. In other systems, the exsolution lamellae can only be observed with the petrographic microscope.

Graphical Representation of Mineral Composition

For simple compositional variation it is often convenient to visualize the compositions in some kind of graphical form. Most chemical analyses of oxide and silicate minerals are reported in weight % oxide components. In weight percent because the classical technique of chemical analyses was once gravimetric, and in oxide components because it is difficult to obtain concentrations of Oxygen, so in oxides and silicates it is assumed that there is enough Oxygen to balance the cationic charges.

If the composition of a mineral can be expressed with 2 components, then a linear scale can be used as a graphical representation of composition. For example, chemical analysis of the mineral kyanite shows that it is composed of about 36 weight % SiO_2 and 64 weight % Al_2O_3. We can plot this on a linear scale as shown here in the upper diagram. In this two component compositional diagram 0% plots at the same point as 100 %.

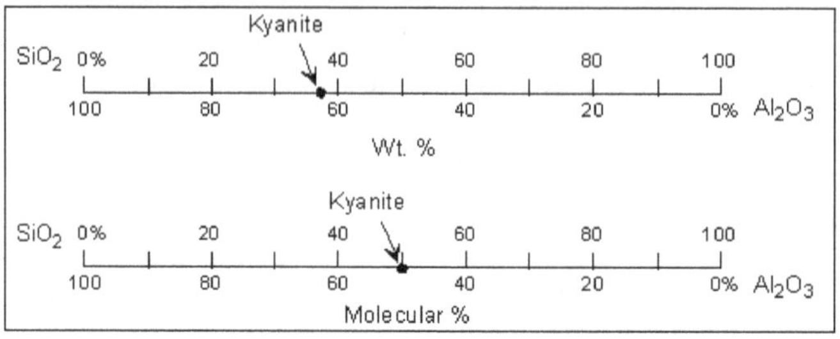

We could divide each of these weight percentages by the molecular weight of each of the oxides and recalculate in the analysis in terms of molecular %. But the chemical formula of Kyanite - Al_2SiO_5 - tells us that kyanite is made up of $1Al_2O_3 + 1SiO_2$. So in molecular percent 50% of kyanite is Al_2O_3 and 50% is SiO_2, as shown in the lower diagram above.

Similarly, olivine can be thought of as a mixture of forsterite, Fo (Mg_2SiO_4) and fayalite, Fa (Fe_2SiO_4). If these are the only two components involved, then note again that 100% Fo corresponds to 0% Fa, and vice versa.

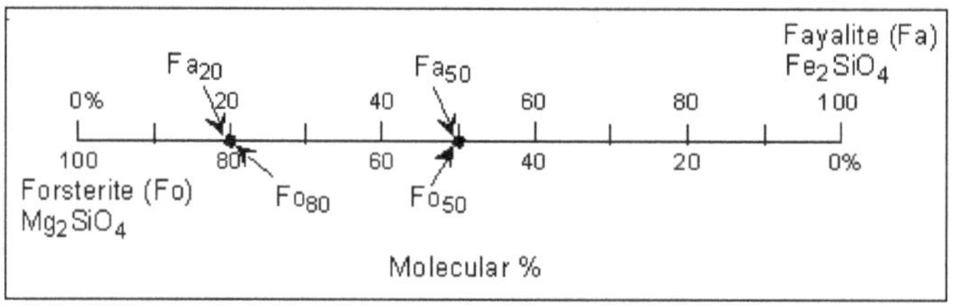

An olivine solid solution that has 50% of the Mg^{+2} ions replaced by Fe^{+2} ions would be said to have a composition Fo_{50} or Fa_{50}. It could be expressed either way, because both ways indicate the same composition. The chemical formula for such a composition would be written as $MgFeSiO_4$. Similarly, for an olivine composition where 20% of the Fe^{+2} ions are replaced by Mg^{+2} ions, the composition could be expressed as Fa_{20} or Fo_{80}.

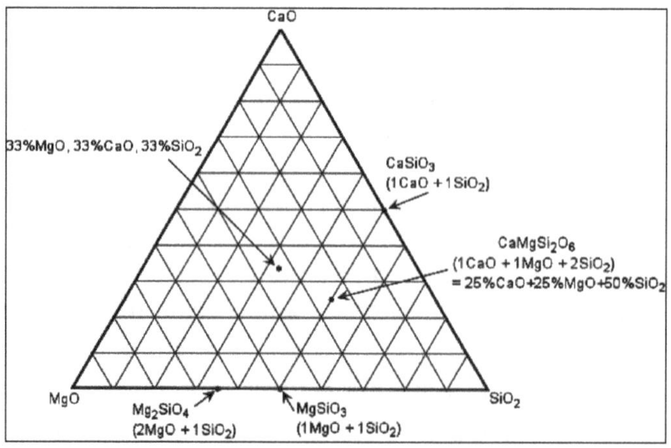

If there are three components that need to be plotted, a triangular graph can be used. Such a graph for the three components MgO, CaO, SiO_2 is shown below. Each of the corners of the triangular graph represent 100% of the component plotted at that corner, and 0% of the other two components. Lines parallel to the sides of the triangle in this case are marked off in 10% increments, so that the horizontals lines represent the % of CaO starting from 0% at the bottom to 100% at the CaO corner. Lines parallel to the SiO_2 – CaO side of the triangle represent the %MgO starting from 0% at the SiO_2 - CaO join to 100% MgO at the MgO corner. Lines parallel to the MgO - CaO side of the diagram represent the % SiO_2.

Note that the composition 33%MgO, 33%CaO, 33%SiO_2 plots at the exact center of the triangle. Minerals that only contain 2 of the three components plot along the sides of the triangle, with the scale being similar to the 2-component graphs. So, for example if we are using molecular percentages:

- $CaSiO_3$ (wollastonite) which can also be written as 1CaO + 1SiO_2, plots 50% of the way between CaO and SiO_2.
- $MgSiO_3$ (enstatite) can also be written as 1MgO + 1SiO_2, and plots 50% of the way between MgO and SiO_2.
- $CaMgSi_2O_6$ (diopside) can also be written as 1CaO + 1MgO +2SiO_2 There are a total of 4 molecules, with 1/4 as CaO, 1/4 as MgO, and 2/4 as SiO_2. So diopside plots at 25% CaO, 25%MgO, and 50%SiO_2.

Triangular diagrams are often used to show the compositional ranges in minerals. We here look at 2 examples. The feldspars can be looked at in terms of the three components Albite (Ab) - $NaAlSi_3O_8$. Orthoclase (Or) - $KALSi_3O_8$, and Anorthite (An) - $CaAl2Si_2O_8$. At high temperature, complete solid solution exists between Ab and Or, to form the alkali feldspar solid solution series. But, as shown in the diagram, the alkali feldspar solid solutions can contain up to 5% of the An component. Similarly a complete solid solution series exists between Ab and an, to form the plagioclase solid solution series. Plagioclase can contain up to about 5% of the or component.

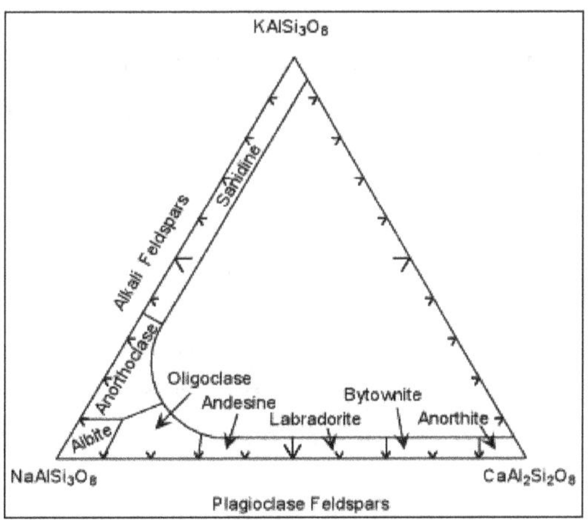

Another example is shown by the pyroxene minerals. These plot in the three component system Enstatite (En) - $MgSiO_3$, Wollastonite (Wo) - $CaSiO_3$, Ferrosilite (Fs) - $FeSiO_3$. Wo does not have a pyroxene structure Complete solid solution exists between Diopside, Di, ($CaMgSi_2O_6$) and Hedenbergite, Hd, ($CaFeSi_2O_6$). These pyroxenes are monoclinic and are thus called the clinopyroxenes.

Augite is also a clinopyroxene, but note that it is depleted in the Wo component relative to the Di – Hd series. En - Fs also forms a complete solid solution series. These minerals are orthorhombic, so the series is often referred to the orthopyroxenes. Pigeonite is a monoclinic pyroxene that has slightly more of the Wo component than the orthopyroxenes.

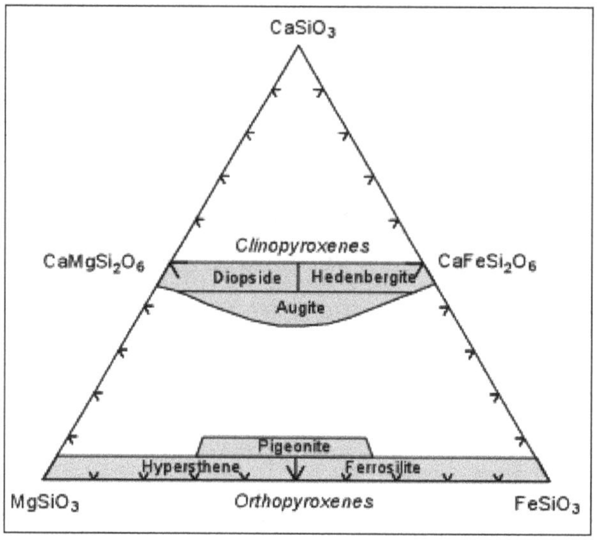

Chemical Properties of Minerals

Every substance on earth is made of atoms, the building blocks of all matter. There are all different types of atoms, and each type of atom is classified as an element. Atoms will bond together to form molecules, which are organized agglomerations of different types of atoms. There are ninety two elements that occur naturally on earth. About twenty of the elements make up over 95 percent of all matter on earth, the others are all rare.

Molecules form mixtures and compounds. Mixtures are groups of molecules that can be separated by physical means. Compounds are groups of molecules with a definitive arrangement, and can only be separated by chemical means.

Every mineral contains a defined ratio of specific molecules in its structure. For example, a water molecule is composed of H_2O - two hydrogen atoms and one oxygen atom. When water molecules are grouped together, they form a body of water. Trace amounts of foreign molecules not inherent to a mineral's makeup are known as impurities. Impurities can slightly alter physical properties such as color. Atoms join together based on their positive and negative charges. This is caused by the amounts of protons (positive charges) or electrons (negative charges) they contain.

Chemical Formula

Every mineral has a unique arrangement of elements within its inherent structure. This arrangement of atoms determines a mineral type. All minerals have a chemical formula, which is an analysis of the types and amounts of elements present in a mineral. Every element has a one or two letter abbreviated term. For example oxygen is "O", and gold is "Au".

The chemical formula of the mineral Hematite is Fe_2O_3. The letters describe the element type (Fe = iron, O = oxygen), and the subscripted numbers describe the amount of those atoms in each molecule. A Hematite molecule has 2 iron (Fe) atoms and 3 oxygen (O) atoms. If there is no number written after an element, it means there is only one atom of that element present.

Radicals

Radicals, or polyatomic ions, are special types of compounds. They act as if they were a single element when they join other elements to form molecules. There are many radicals, some of the most familiar are the:

Carbonate radical	CO_3
Sulfate Radical	SO_4
Chromate Radical	CrO_4
Hydroxyl Radical	OH

Radicals are treated like any element when written in a chemical formula. Radicals are sometimes surrounded with a parenthesis when written in a chemical formula, and the number after the parenthesis describes how many of these radicals are present within each mineral's molecular breakdown. For example, the mineral Talc has a chemical formula of $Mg_3Si_4O_{10}(OH)_2$. The subscripted 2 at the end of the (OH) describes that there are two hydroxyl radicals within each molecule of Talc. If there is no subscripted number written after a radical, that means there is only one radical present in that mineral's molecule.

Hydrous Minerals

Minerals containing water in their structure are known as hydrous minerals. The hydrous mineral Gypsum has a chemical formula of "$CaSO_4 \cdot 2H_2O$". The large number 2 in front of the H_2O signifies that there are two water (H_2O) molecules for every molecule of $CaSO_4$. The dot in between $CaSO_4$ and $2H_2O$ indicates that these are two separate molecules, but they are rationally proportionate.

The letter "n" is used to describe a variable amount of water in the structure of a mineral. For example, the hydrous mineral Opal has a loosely defined composition with an inconsistent amount of water in its structure, thus its chemical formula is written as "$SiO_2 \cdot nH_2O$".

Some minerals, such as Torbernite, have a varying amount of water within fixed limits. The chemical formula for Torbernite is "$Cu(UO_2)(PO_4)_2 \cdot 8\text{-}12H_2O$". The $8\text{-}12H_2O$ indicates that there can be 8 to 12 water molecules for every $Cu(UO_2)(PO_4)2$ molecule.

Mineral Series

A number of minerals contain a varying amount of two or more elements. For example, the mineral Aurichalcite, which has a chemical formula of $(Zn,Cu)_5(CO_3)_2(OH)_6$ contains an unspecific varying amount of zinc (Zn) and copper (Cu). This is indicated by comma separating the Zn from the Cu.

If a chemical formula with two elements in parenthesis is separated by a comma, the numbers of those elements vary. Aurichalcite has a variable amount of zinc of copper where the combination of both these elements totals five. The more dominant element is usually listed first.

Often, when the elements in a mineral vary, a series is formed. A series consists of a group of mineral in which one of the elements varies. For example, the Spinel series contains four members, with a series formula of $(Mg,Zn,Fe,Mn)Al_2O_4$. There are four end members of this series, which are listed below. Intermediary forms that are a combination of two or more also exist.

Spinel End Members		
Spinel	$MgAl_2O_4$	Magnesium Spinel
Gahnite	$ZnAl_2O_4$	Zinc Spinel
Hercynite	$FeAl_2O_4$	Iron Spinel
Galaxite	$MnAl_2O_4$	Manganese Spinel

Some intermediary members of the spinel group have designated names, such as Gahnospinel $(Mg,Zn)Al_2O_4$, which is a mixture of Spinel ($MgAl_2O_4$) and Gahnite ($ZnAl_2O_4$); while other mixtures lack a designated name and are just called under the umbrella of Spinel. Intermediary members are sometimes scientifically recognized as individual mineral species by the IMA; though they are more often not recognized (Gahnospinel is not recognized as a scientifically distinct mineral by the IMA).

Many mineral series form solid solutions. In a solid solution, there are intermediate members between the two end members. Individual intermediary members may be given names, while others may not. An example is the Olivine group, with a chemical formula of $(Mg,Fe)_2SiO_4$. It contains Forsterite, (Mg_2SiO_4), and Fayalite, (Fe_2SiO_4) as the two end members. Very rarely are the Olivine members pure; most are somewhere along the intermediary scale within the solid solution series, with a varying percentage of Mg and Fe.

Atomical Variations

There are certain elements that come in slight modifications. Some elements have different amounts of electrons in atoms of the same element. Some minerals, such as Babingtonite, contain two types of the same element. The chemical formula for Babingtonite is written $Ca2Fe_2+Fe_3+SiO1_4(OH)$. The superscript form of 2^+ and 3^+ next to the iron (Fe) distinguishes the different types of iron. 2^+ means the iron has 2 more electrons than protons, and 3^+ means it has three more electrons. The amount of electrons affects the chemical bonding of any element with variable electrons. Anytime there is more than one variable atom in a molecule, a superscript number representing the type of atom (i.e. how many electrons over protons) is written after the element symbol.

Chemical Formula Variables

The chemical formula of some minerals may be written in different formats and styles. It is not uncommon to see chemical formulas written differently for the same mineral. For example, the chemical formula for the mineral Dioptase is usually written $CuSiO_2\ (OH)_2$. However, its formula may also be written H_2CuSiO_4. Both formulas mean the exact same thing, as there is an equal amount of all the atoms in both formulas.

Sometimes a formula may be reduced or expanded by using multiplication or division. (A reduced formula is known as an empirical formula in chemistry.) For example, the chemical formula of Sodalite is commonly written as $Na_4Al_3Si_3O_{12}Cl$. The silicon and oxygen form a radical, so they are bound together as a single atom. Since the amount of silicon and oxygen is divisible by three ($Si_3 \div_3 O_{12} \div_3$ = SiO_4), the chemical formula can also be written as $Na_4Al_3(SiO_4)_3Cl$.

Variable Formulas

Some minerals may contain an element that partially replaces an inherent element. For example, the mineral Adamite, Zn_2 (AsO_4) (OH), often contains small amounts of Cu (copper) and Co (cobalt) replacing some Zn (zinc). These elements are not mentioned in the chemical formula, as they do not compromise a significant portion and are only occasionally present.

The variable formula displays the regular formula with the additional elements that are occasionally present. The inherent element is listed first in the parenthesis, and the occasional replacement elements are listed after and underlined. For example, the variable formula of Adamite is $(Zn,Cu,Co)_2(AsO_4)(OH)$. The Zn (zinc) is always present, and the Cu and Co may be present in small amounts or may not be.

Some minerals without a variable formula may still have variable elements, but those elements are too rare or insignificant to be reckoned. For example, the mineral Fluorite occasionally has contains traces of Ce (cerium) and Y (yttrium), but there is no variable formula since these combinations are too uncommon.

Chemical Weathering of Minerals

Chemical weathering results from chemical changes to minerals that become unstable when they are exposed to surface conditions. The kinds of changes that take place are highly specific to the mineral and the environmental conditions. Some minerals, like quartz, are virtually unaffected by chemical weathering, while others, like feldspar, are easily altered. In general, the degree of chemical weathering is greatest in warm and wet climates, and least in cold and dry climates. The important characteristics of surface conditions that lead to chemical weathering are the presence of water (in the air and on the ground surface), the abundance of oxygen, and the presence of carbon dioxide, which produces weak carbonic acid when combined with water. That process, which is fundamental to most chemical weathering, can be shown as follows:

$$H_2O + CO_2 \longrightarrow H_2CO_3 \text{ then } H_2CO_3 \longrightarrow H^+ + HCO_3^-,$$

water + carbon dioxide ⟶ carbonic acid then carbonic acid ⟶ hydrgen ion + carbonate ion

Here we have water (e.g., as rain) plus carbon dioxide in the atmosphere, combining to create carbonic acid. Then carbonic acid dissociates (comes apart) to form hydrogen and carbonate ions. The amount of CO_2 in the air is enough to make only very weak carbonic acid, but there is typically much more CO_2 in the soil, so water that percolates through the soil can become significantly more acidic.

There are two main types of chemical weathering. On the one hand, some minerals become altered to other minerals. For example, feldspar is altered — by hydrolysis — to clay minerals. On the other hand, some minerals dissolve completely, and their components go into solution. For example, calcite ($CaCO_3$) is soluble in acidic solutions.

The hydrolysis of feldspar can be written like this:

$$CaAl_2Si_2O_8 + H_2CO_3 + \frac{1}{2}O_2 \longrightarrow Al_2Si_2O_5(OH)_4 + Ca^{2+} + CO_3^{2-}$$

Plagioclase + carbonic acid ⟶ kaolinite + dissolved calcium + carbonate ions

This reaction shows calcium plagioclase feldspar, but similar reactions could also be written for sodium or potassium feldspars. In this case, we end up with the mineral kaolinite, along with calcium and carbonate ions in solution. Those ions can eventually combine (probably in the ocean) to form the mineral calcite. The hydrolysis of feldspar to clay, which shows two images of the same granitic rock, a recently broken fresh surface on the left and a clay-altered weathered surface on the right. Other silicate minerals can also go through hydrolysis, although the end results will be a little different. For example, pyroxene can be converted to the clay minerals chlorite or smectite, and olivine can be converted to the clay mineral serpentine.

Figure: Unweathered (left) and weathered (right) surfaces of the same piece of granitic rock. On the unweathered surfaces the feldspars are still fresh and glassy-looking. On the weathered surface the feldspar has been altered to the chalky-looking clay mineral kaolinite.

Oxidation is another very important chemical weathering process. The oxidation of the iron in a ferromagnesian silicate starts with the dissolution of the iron. For olivine, the process looks like this, where olivine in the presence of carbonic acid is converted to dissolved iron, carbonate, and silicic acid:

$$Fe_2SiO_4 + 4H_2CO_3 \longrightarrow 2Fe^{2+} + 4HCO_3^- + H_4SiO_4$$

Olivine + (carbonic acid) ⟶ dissolved iron + dissolved carbonate + dissolved silicic acid In the presence of oxygen, the dissolved iron is then quickly converted to hematite:

$$2Fe_2^+ + 4HCO_3^- + \frac{1}{2}O_2 + 2H_2O \longrightarrow Fe_2O_3 + 4H_2CO_3$$

Dissolved iron + bicarbonate + oxygen + water ⟶hematite + carbonic acid. The equation for olivine, but it could apply to almost any other ferromagnesian silicate, including pyroxene, amphibole, or biotite. Iron in the sulphide minerals (e.g., pyrite) can also be oxidized in this way. And the mineral hematite is not the only possible end result, as there is a wide range of iron oxide minerals that can form in this way. The results of this process, which shows a granitic rock in which some of the biotite and amphibole have been altered to form the iron oxide mineral limonite.

Figure: A granitic rock containing biotite and amphibole which have been altered near to the rock's surface to limonite, which is a mixture of iron oxide minerals.

A special type of oxidation takes place in areas where the rocks have elevated levels of sulphide minerals, especially pyrite (FeS_2). Pyrite reacts with water and oxygen to form sulphuric acid, as follows:

$$2FeS_2 + 7O_2 + 2H_2O \longrightarrow 2Fe^{2+} H_2SO_4 + 2H^+$$

pyrite + oxygen + water ⟶ iron ions + sulphuric acid + hydrogen ions

The runoff from areas where this process is taking place is known as acid rock drainage (ARD), and even a rock with 1% or 2% pyrite can produce significant ARD. Some of the worst examples of ARD are at metal mine sites, especially where pyrite-bearing rock and waste material have been mined from deep underground and then piled up and left exposed to water and oxygen. One example of that is the Mt. Washington Mine near Courtenay on Vancouver Island, but there are many similar sites across Canada and around the world.

Figure: Exposed oxidizing and acid generating rocks and mine waste at the abandoned Mt. Washington Mine, B.C. (left), and an example of acid drainage downstream from the mine site (right).

At many ARD sites, the pH of the runoff water is less than 4 (very acidic). Under these conditions, metals such as copper, zinc, and lead are quite soluble, this can lead to toxicity for aquatic and other organisms. For many years, the river downstream from the Mt. Washington Mine had so much dissolved copper in it that it was toxic to salmon. Remediation work has since been carried out at the mine and the situation has improved.

The hydrolysis of feldspar and other silicate minerals and the oxidation of iron in ferromagnesian silicates all serve to create rocks that are softer and weaker than they were to begin with and thus more susceptible to mechanical weathering.

The weathering reactions so far involved the transformation of one mineral to another mineral (e.g., feldspar to clay), and the release of some ions in solution (e.g., Ca^{2+}). Some weathering processes involve the complete dissolution of a mineral. Calcite, for example, will dissolve in weak acid, to produce calcium and bicarbonate ions. The equation is as follows:

$$CaCO_3 + H^+ + HCO_3^- \longrightarrow Ca^{2+} + 2HCO_3^-$$

calcite + hydrogen ions + bicarbonate ⟶ calcium ions + bicarbonate

Calcite is the major component of limestone (typically more than 95%), and under surface conditions, limestone will dissolve to varying degrees (depending on which minerals it contains, other than calcite). Limestone also dissolves at relatively shallow depths underground, forming limestone caves.

Figure: A limestone outcrop. The limestone, which is primarily made up of the mineral calcite, has been dissolved to different degrees in different areas because of compositional differences. The buff-coloured bands are volcanic rock, which is not soluble.

Metamorphism

The mineralogical and structural adjustment of solid rocks to physical and chemical conditions that have been imposed at depths below the near surface zones of weathering and diagenesis and which differ from conditions under which the rocks in question originated.

In geology this refers to the changes in mineral assemblage and texture that result from subjecting a rock to conditions such pressures, temperatures, and chemical environments different from those under which the rock originally formed:

- Diagenesis is also a change in form that occurs in sedimentary rocks. In geology, however, we restrict diagenetic processes to those which occur at temperatures below 200°C and pressures below about 300 MPa (MPa stands for Mega Pascals); this is equivalent to about 3 kilobars of pressure (1kb = 100 MPa).
- Metamorphism therefore occurs at temperatures and pressures higher than 200°C and 300 MPa. Rocks can be subjected to these higher temperatures and pressures as they are buried deeper in the Earth. Such burial usually takes place as a result of tectonic processes such as continental collisions or subduction.

- The upper limit of metamorphism occurs at the pressure and temperature where melting of the rock in question begins. Once melting begins the process changes to an igneous process rather than a metamorphic process.

Grade of Metamorphism

As the temperature and/or pressure increases on a body of rock we say the rock undergoes prograde metamorphism or that the grade of metamorphism increases. Metamorphic grade is a general term for describing the relative temperature and pressure conditions under which metamor-phic rocks form:

- Low-grade metamorphism takes place at temperatures between about 200 to 320°C, and relatively low pressure. Low grade metamorphic rocks are generally characterized by an abundance of hydrous minerals. With increasing grade of metamorphism, the hydrous minerals begin to react with other minerals and/or break down to less hydrous minerals.

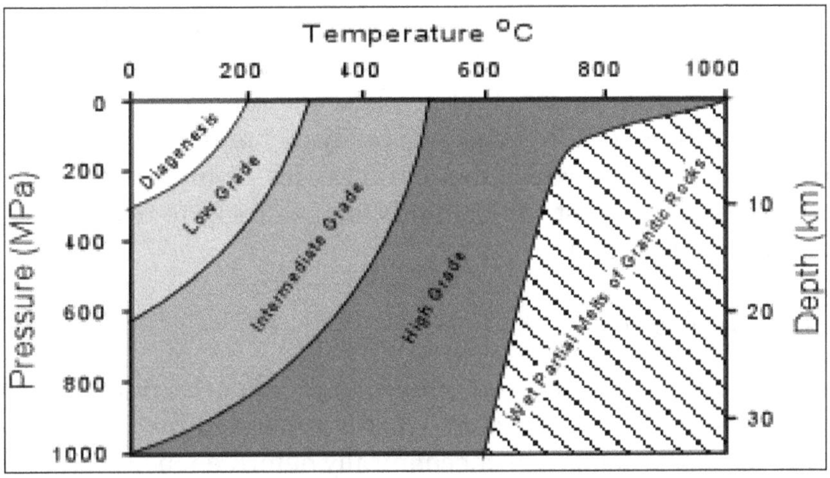

- High-grade metamorphism takes place at temperatures greater than 320°C and relatively high pressure. As grade of metamorphism increases, hydrous minerals become less hydrous, by losing H_2O, and non-hydrous minerals become more common.

Types of Metamorphism

Contact Metamorphism

Contact metamorphism occurs adjacent to igneous intrusions and results from high temperatures associated with the igneous intrusion. Since only a small area surrounding the intrusion is heated by the magma, metamorphism is restricted to the zone surrounding the intrusion, called a metamorphic or contact aureole. Outside of the contact aureole, the rocks are not affected by the intrusive event. The grade of metamorphism increases in all directions toward the intrusion. Because the temperature contrast between the surrounding rock and the intruded magma is larger at shallow levels in the crust where pressure is low, contact metamorphism is often referred to as high temperature, low pressure metamorphism. The rock produced is often a fine-grained rock that shows no foliation, called a hornfels.

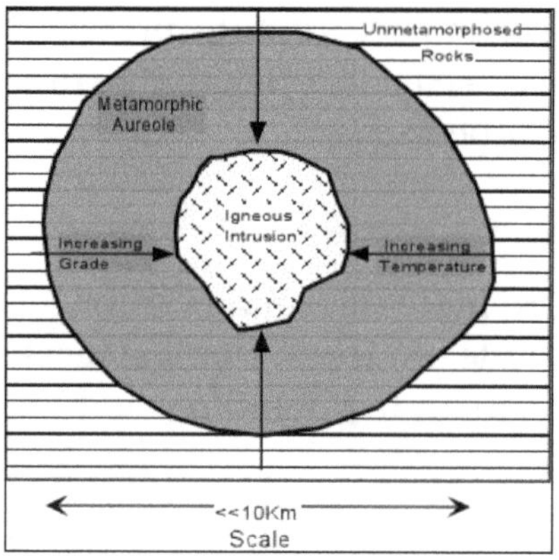

Regional Metamorphism

Regional metamorphism occurs over large areas and generally does not show any relationship to igneous bodies. Most regional metamorphism is accompanied by deformation under non-hydrostatic or differential stress conditions. Thus, regional metamorphism usually results in forming metamorphic rocks that are strongly foliated, such as slates, schists, and gniesses. The differential stress usually results from tectonic forces that produce compressional stresses in the rocks, such as when two continental masses collide. Thus, regionally metamorphosed rocks occur in the cores of fold/thrust mountain belts or in eroded mountain ranges. Compressive stresses result in folding of rock and thickening of the crust, which tends to push rocks to deeper levels where they are subjected to higher temperatures and pressures.

Cataclastic Metamorphism

Cataclastic metamorphism occurs as a result of mechanical deformation, like when two bodies of rock slide past one another along a fault zone. Heat is generated by the friction of sliding along such a shear zone, and the rocks tend to be mechanically deformed, being crushed and pulverized, due to the shearing. Cataclastic metamorphism is not very common and is restricted to a narrow zone along which the shearing occurred.

Hydrothermal Metamorphism

Rocks that are altered at high temperatures and moderate pressures by hydrothermal fluids are hydrothermally metamorphosed. This is common in basaltic rocks that generally lack hydrous minerals. The hydrothermal metamorphism results in alteration to such Mg-Fe rich hydrous minerals as talc, chlorite, serpentine, actinolite, tremolite, zeolites, and clay minerals. Rich ore deposits are often formed as a result of hydrothermal metamorphism.

Burial Metamorphism

When sedimentary rocks are buried to depths of several kilometers, temperatures greater than 300°C may develop in the absence of differential stress. New minerals grow, but the rock does not appear to be metamorphosed. The main minerals produced are often the Zeolites. Burial metamorphism overlaps, to some extent, with diagenesis, and grades into regional metamorphism as temperature and pressure increase.

Shock Metamorphism (Impact Metamorphism)

When an extraterrestrial body, such as a meteorite or comet impacts with the Earth or if there is a very large volcanic explosion, ultrahigh pressures can be generated in the impacted rock. These ultrahigh pressures can produce minerals that are only stable at very high pressure, such as the SiO_2 polymorphs coesite and stishovite. In addition they can produce textures known as shock lamellae in mineral grains, and such textures as shatter cones in the impacted rock.

Classification of Metamorphic Rocks

Classification of metamorphic rocks is based on mineral assemblage, texture, protolith, and bulk chemical composition of the rock.

Texture

In metamorphic rocks individual minerals may or may not be bounded by crystal faces. Those that are bounded by their own crystal faces are termed idioblastic. Those that show none of their own crystal faces are termed xenoblastic. From examination of metamorphic rocks, it has been found that metamorphic minerals can be listed in a generalized sequence, known as the crystalloblastic series, listing minerals in order of their tendency to be idioblastic. In the series, each mineral tends to develop idioblastic surfaces against any mineral that occurs lower in the series. This series is listed below:

- Rutile, sphene, magnetite.
- Tourmaline kyanite, staurolite, garnet, andalusite.
- Epidote, zoisite, lawsonite, forsterite.
- Pyroxenes, amphiboles, wollastonite.
- Micas, chlorites, talc, stilpnomelane, prehnite.
- Dolomite, calcite.
- Scapolite, cordierite, feldspars.
- Quartz.

This series can, in a rather general way, enable us to determine the origin of a given rock. For example a rock that shows euhedral plagioclase crystals in contact with anhedral amphibole, likely had an igneous protolith, since a metamorphic rock with the same minerals would be expected to show euhedral amphibole in contact with anhedral plagioclase.

Another aspect of the crystalloblastic series is that minerals high on the list tend to form porphyroblasts (the metamorphic equivalent of phenocrysts), although K-feldspar (a mineral that occurs lower in the list) may also form porphyroblasts. Porphyroblasts are often riddled with inclusions of other minerals that were enveloped during growth of the porphyroblast. These are said to have a poikioblastic texture.

Most metamorphic textures involve foliation. Foliation is generally caused by a preferred orientation of sheet silicates. If a rock has a slatey cleavage as its foliation, it is termed a slate, if it has a phyllitic foliation, it is termed a phyllite, if it has a shistose foliation, it is termed schist. A rock that shows a banded texture without a distinct foliation is termed gneiss. All of these could be porphyroblastic (i.e. could contain porhyroblasts). A rock that shows no foliation is called a hornfels if the grain size is small, and a granulite, if the grain size is large and individual minerals can be easily distinguished with a hand lens.

Protolith

Protolith refers to the original rock, prior to metamorphism. In low grade metamorphic rocks, original textures are often preserved allowing one to determine the likely protolith. As the grade of metamorphism increases, original textures are replaced with metamorphic textures and other clues, such as bulk chemical composition of the rock, are used to determine the protolith.

Bulk Chemical Composition

- The pressure and temperature reached during metamorphism.
- The composition of any fluid phase present during metamorphism.
- The bulk chemical composition of the rock.

Just like in igneous rocks, minerals can only form if the necessary chemical constituents are present in the rock (i.e. the concept of silica saturation and alumina saturation applies to metamorphic rocks as well). Based on the mineral assemblage present in the rock one can often estimate the approximate bulk chemical composition of the rock. Some terms that describe this general bulk chemical composition are as follows:

- Pelitic: These rocks are derivatives of aluminous sedimentary rocks like shales and mudrocks. Because of their high concentrations of alumina they are recognized by an abundance of aluminous minerals, like clay minerals, micas, kyanite, sillimanite, andalusite, and garnet.
- Quartzo-Feldspathic: Rocks that originally contained mostly quartz and feldspar like granitic rocks and arkosic sandstones will also contain an abundance of quartz and feldspar as metamorphic rocks, since these minerals are stable over a wide range of temperature and pressure. Those that exhibit mostly quartz and feldspar with only minor amounts of aluminous minerals are termed quartzo-feldspathic.
- Calcareous: Calcareous rocks are calcium rich. They are usually derivatives of carbonate rocks, although they contain other minerals that result from reaction of the carbonates with associated siliceous detrital minerals that were present in the rock. At low grades of metamorphism calcareous rocks are recognized by their abundance of carbonate minerals like calcite and dolomite. With increasing grade of metamorphism these are replaced by minerals like brucite, phlogopite (Mg-rich biotite), chlorite, and tremolite. At even higher grades anhydrous minerals like diopside, forsterite, wollastonite, grossularite, and calcic plagioclase.
- Basic: Just like in igneous rocks, the general term basic refers to low silica content. Basic metamorphic rocks are generally derivatives of basic igneous rocks like basalts and

gabbros. They have an abundance of Fe-Mg minerals like biotite, chlorite, and hornblende, as well as calcic minerals like plagioclase and epidote.

- Magnesian: Rocks that are rich in Mg with relatively less Fe, are termed magnesian. Such rocks would contain Mg-rich minerals like serpentine, brucite, talc, dolomite, and tremolite. In general, such rocks usually have an ultrabasic protolith, like peridotite, dunite, or pyroxenite.
- Ferriginous: Rocks that are rich in Fe with little Mg are termed ferriginous. Such rocks could be derivatives of Fe-rich cherts or ironstones. They are characterized by an abundance of Fe-rich minerals like greenalite (Fe-rich serpentine), minnesotaite (Fe-rich talc), ferroactinolite, ferrocummingtonite, hematite, and magnetite at low grades, and ferrosilite, fayalite, ferrohedenbergite, and almandine garnet at higher grades.
- Manganiferrous: Rocks that are characterized by the presence of Mn-rich minerals are termed manganiferrous. They are characterized by such minerals as Stilpnomelane and spessartine.

Classification

Classification of metamorphic rocks depends on what is visible in the rock and its degree of metamorphism. Note that classification is generally loose and practical such that names can be adapted to describe the rock in the most satisfactory way that conveys the important characteristics. Three kinds of criteria are normally employed. These are:

- Mineralogical: The most distinguishing minerals are used as a prefix to a textural term. Thus, a schist containing biotite, garnet, quartz, and feldspar, would be called biotite-garnet schist. A gneiss containing hornblende, pyroxene, quartz, and feldspar would be called hornblende-pyroxene gneiss. A schist containing porphyroblasts of K-feldspar would be called K-spar porphyroblastic schist.
- Chemical: If the general chemical composition can be determined from the mineral assemblage, then a chemical name can be employed. For example schist with a lot of quartz and feldspar and some garnet and muscovite would be called garnet-muscovite quartzo-feldspathic schist. Schist consisting mostly of talc would be called talc-magnesian schist.
- Protolithic: If a rock has undergone only slight metamorphism such that its original texture can still be observed then the rock is given a name based on its original name, with the prefix meta- applied. For example: metabasalt, metagraywacke, meta-andesite, metagranite.

In addition to these conventions, certain non-foliated rocks with specific chemical compositions and/or mineral assemblages are given specific names. These are as follows:

- Amphibolites: These are medium to coarse grained, dark colored rocks whose principal minerals are hornblende and plagioclase. They result from metamorphism of basic igneous rocks. Foliation is highly variable, but when present the term schist can be appended to the name (i.e. amphibolite schist).
- Marbles: These are rocks composed mostly of calcite, and less commonly of dolomite. They result from metamorphism of limestones and dolostones. Some foliation may be present if the marble contains micas.

- Eclogites: These are medium to coarse grained consisting mostly of garnet and green clinopyroxene called omphacite, that result from high grade metamorphism of basic igneous rocks. Eclogites usually do not show foliation.
- Quartzites: Quartz arenites and chert both are composed mostly of SiO_2. Since quartz is stable over a wide range of pressures and temperatures, metamorphism of quartz arenites and cherts will result only in the recrystallization of quartz forming a hard rock with interlocking crystals of quartz. Such a rock is called a quartzite.
- Serpentinites: Serpentinites are rocks that consist mostly of serpentine. These form by hydrothermal metamorphism of ultrabasic igneous rocks.
- Soapstones: Soapstones are rocks that contain an abundance of talc, which gives the rock a greasy feel, similar to that of soap. Talc is an Mg-rich mineral, and thus soapstones from ultrabasic igneous protoliths, like peridotites, dunites, and pyroxenites, usually by hydrothermal alteration.
- Skarns: Skarns are rocks that originate from contact metamorphism of limestones or dolostones, and show evidence of having exchanged constituents with the intruding magma. Thus, skarns are generally composed of minerals like calcite and dolomite, from the original carbonate rock, but contain abundant calcium and magnesium silicate minerals like andradite, grossularite, epidote, vesuvianite, diopside, and wollastonite that form by reaction of the original carbonate minerals with silica from the magma. The chemical exchange is that takes place is called metasomatism.
- Mylonites: Mylonites are cataclastic metamorphic rocks that are produced along shear zones deep in the crust. They are usually fine-grained, sometimes glassy, that are streaky or layered, with the layers and streaks having been drawn out by ductile shear.

Metamorphic Facies

In general, metamorphic rocks do not drastically change chemical composition during metamorphism, except in the special case where metasomatism is involved (such as in the production of skarns). The changes in mineral assemblages are due to changes in the temperature and pressure conditions of metamorphism. Thus, the mineral assemblages that are observed must be an indication of the temperature and pressure environment that the rock was subjected to. This pressure and temperature environment is referred to as Metamorphic Facies. (This is similar to the concept of sedimentary facies, in that a sedimentary facies is also a set of environmental conditions present during deposition). The sequence of metamorphic facies observed in any metamorphic terrain, depends on the geothermal gradient that was present during metamorphism.

A high geothermal gradient such as the one labeled "A", might be present around an igneous intrusion, and would result in metamorphic rocks belonging to the hornfels facies. Under a normal to high geothermal gradient, such as "B", rocks would progress from zeolite facies to greenschist, amphibolite, and eclogite facies as the grade of metamorphism (or depth of burial) increased. If a low geothermal gradient was present, such the one labeled "C" in the diagram, then rocks would progress from zeolite facies to blueschist facies to eclogite facies.

Thus, if we know the facies of metamorphic rocks in the region, we can determine what the geothermal gradient must have been like at the time the metamorphism occurred. This relationship between geothermal gradient and metamorphism will be the central theme of our discussion of metamorphism.

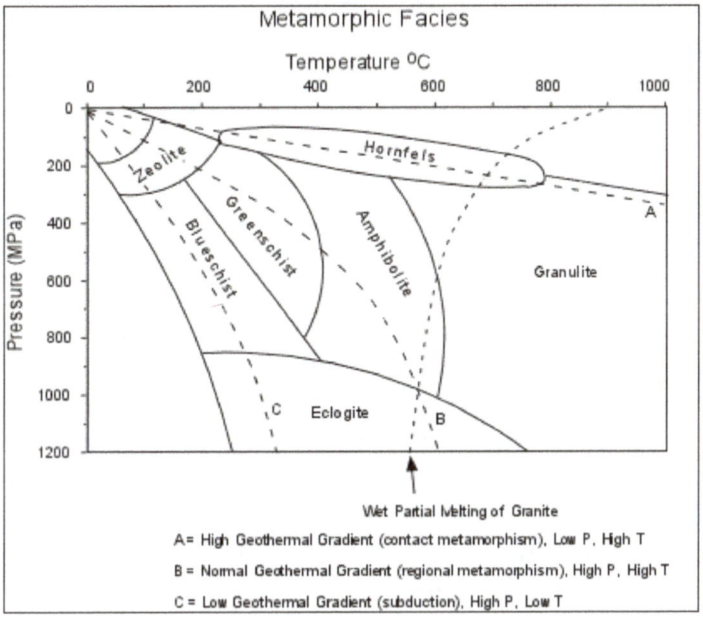

Metasomatism

Metasomatism refers to one mineral replacing another mineral phenomenon. Its essence is that the dissolution of minerals and metasomatic minerals deposition occurs at the same time, leading to the metasomatism occurrence. This effect often occurs along vesicles, intergranular pores, and cracks and has typical metasomatic stripes characteristics. Volcanic rocks metasomatism refers to the replacement of various porphyries and crystals replacement with albite, amphibole, siderite, quartz, carbonate, gypsum, kaolin and chlorite, etc. It occurs from the hydrothermal stage to the burial stage. The metasomatism effect on primary pores is related to the action type. For example, the pore volume loss due to metasomatism accounts for less than 2% of the total pore volume in the XX gas field volcanic reservoirs.

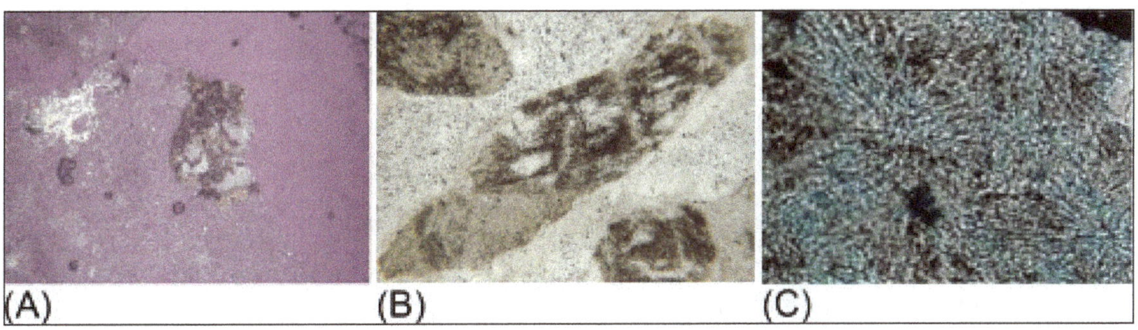

Figure: Metasomatism and recrystallization of secondary minerals: (A) calcite replacing quartz (XX2, 3952 m, felsite), (B) feldspar phenocryst being replaced by albite (XX9, 3905 m, dacite), and (C) recrystallization (XX901, 3864.22 m, rhyolitic tuff lava).

Recrystallization refers to the mineral recrystallization effect after consolidation, or the process during which dissolution, component migration, and precipitation occur again on the same mineral in the solid state without producing new mineral phases. Volcanic rock recrystallization mainly refers to the matrix recrystallization process, forming a spherulite texture or felsitic texture. This effect mostly occurs from the hydrothermal stage to the burial stage, which mainly affects the pores variation and matrix micropores. The primary pore volume loss is about 1%.

Most treatments of metasomatism and mineralising processes adopt the concept introduced by Korzhinskii of mosaic equilibrium which has come to be known in the geological literature as local equilibrium. The proposal is that even though the system is not at equilibrium one can identify a large-enough region for the system that can be considered to be at equilibrium. The evolution of the system can then be built up by considering the interaction between these regions. The notion of equilibrium is pervasive in metamorphic and metasomatic petrology, to the extent that the concept is commonly treated as self-evident for most geological systems. By definition, however, systems are not at equilibrium during their evolution and there are many situations, which we will discuss, where even the smallest region, in which concepts such as temperature and internal energy can be defined by some kind of averaging of molecular motions, is not at equilibrium. A self-evident form of this statement is that mineral reactions cannot be at equilibrium while the mineral reactions are in progress and minerals are growing. There must be chemical affinities driving the reactions and hence by definition the system is not at equilibrium; at equilibrium these affinities drop to zero not meant as a threat to the vast amount of work that has been devoted to defining mineral systems at equilibrium. If we knew how to handle systems not at equilibrium it would give us an extra toolbox that enables us to investigate the processes that operate during deformation and metamorphism.

References

- Mineral-chemistry: tulane.edu, Retrieved 25, June 2020
- Chemical-properties, resource: minerals.net, Retrieved 04, March 2020
- Chemical-weathering, geology: opentextbc.ca, Retrieved 17, August 2020
- Typesmetamorph: tulane.edu, Retrieved 09, July 2020
- Metasomatism, earth-and-planetary-sciences: sciencedirect.com, Retrieved 12, January 2020

Chapter 3

Mineral Physics

Mineral physics is the science of geological and geo-inspired materials with emphasis on their structures, properties and behavior. Optical properties, magnetic properties, mineral strength, electrical properties, etc. are aspects which fall under its domain. This chapter closely examines all these concepts associated with mineral physics to provide an extensive understanding of the subject.

Mineral physics is the materials science of geological and geo-inspired synthetic materials, with emphasis on variations of structure, properties, and behavior of minerals, melts, and fluids as a function of pressure and temperature. Our field is transitioning from assembling an immense dossier of information about these Earth materials to elucidating the dynamic response of Earth to planetary-scale driving forces. Answers to many outstanding post-plate tectonics revolution geoscience questions hinge on knowledge of the equilibrium and dynamic properties of rocks, minerals, and melts; for example: How were Earth and the solar system formed? What is Earth's current state in the context of our planetary-scale evolutionary trajectory? What combination of factors has resulted in Earth's distinct geology and habitability?

The field of mineral physics provides ever-improving measurements of and constraints on material behavior that refine interpretation of growing suites of geophysical and geochemical observations. Groundbreaking discoveries driven by technological advances in materials synthesis, experiments, and analysis stimulate development of new hypotheses about the inner workings of Earth and other planets. As a result, mineral physics has grown hand in hand with the geosciences, responsive to the needs of the broader scientific community while simultaneously catalyzing transformative advances.

The same suite of tools used to study Earth materials also successfully addresses some of society's needs for new and improved materials to drive economic development and technological innovation that serve the planet, while striving for energy sustainability. Examples include discovery of new superhard ceramics and improvements to high-temperature superconductors, multiferroics, selective gas absorbers, and photovoltaics.

The goal of the planning workshop held in Argonne, Illinois, on October 10–12, 2014, and this report is to scope the array of promising technological developments and connect them with some of the most compelling questions in the geosciences. Compared to five years ago, we can now reach considerably higher pressures and higher temperatures with larger sample volumes, study more complex and more realistic samples, reliably constrain the compressional behavior of melts and liquids, and access the domain of dynamic effects and strain/stress rates. Many of these innovations have become available only recently or are still in development, thus, their full technological potential has yet to be realized. Making the most of new technologies requires concerted and coordinated planning of research and development efforts, partnered with a crystal-clear focus on framing the most compelling Earth and planetary science questions and providing the sharpest answers.

Breakthroughs in experimental technology and cyberinfrastructure stimulate new research directions and open doors to experiments and computations that we could only dream about a decade ago. Mineral physics has proven very adaptable to technological advances and has been able to quickly and creatively take advantage of these new capabilities. Based on the workshop discussions, several very promising directions were identified as priority areas for technological development: superfast detectors and innovations in pressure-generating devices that enable remote, digital pressure control and modulation according to specific shape functions for use in time-resolved studies of dynamic phenomena; a high-performance, large-volume, multi-anvil press to explore aspects of multicomponent phase equilibria, melt properties, mineral elasticity, trace element partitioning, volatiles, and oxygen fugacity in Earth's lower mantle; microfabrication techniques that offer exceptional opportunities for designing the next generation of experiments with microscale control; and cyberinfrastructure solutions designed for extremeconditions science and mineral physics.

Optical Properties

Optical mineralogy is the study of the interaction of light with minerals, most commonly limited to visible light and usually further limited to the non-opaque minerals. Opaque minerals are more commonly studied in reflected light and that study is generally called ore microscopy - alluding to the fact many opaque minerals are also ore minerals. The most general application of optical mineralogy is to aid in the identification of minerals, either in rock thin sections or individual mineral grains. Another application occurs because the optical properties of minerals are related to the crystal chemistry of the mineral - for example, the mineral's chemical composition, crystal structure, order/disorder. Thus, relationships exist, and correlations are possible between them and some optical property. This often allows a simple optical measurement with the petrographic microscope (a polarizing microscope) that may yield important information about some crystal chemical aspect of the mineral under study.

Light

Theory of Light

Light can be thought of both as a wave phenomenon (electromagnetic theory) or a particle phenomenon (quantum theory), depending upon the physical process under study. In optical mineralogy both forms of light are employed to fully explain the interaction of light with minerals. The range and colors of visible light are defined in terms of wavelength: violet (390-446 nm), indigo (446-464 nm), blue (464-500 nm), green (500-578 nm), yellow (578-592 nm), orange (592-620 nm), and red (620-770 nm). Visible light commonly is referred to as 400-700 nm. This range is somewhat arbitrary based upon the variation in an individual's eye sensitivity.

The wave theory of light describes light as a longitudinal wave, with the direction of propagation and the direction of energy transfer being perpendicular. In optical mineralogy the direction of propagation is referred to as the ray path and the energy transfer direction as the vibration direction. The geometrical relationships between ray path, vibration direction, and a mineral constitute one major portion of the optical study of minerals, basically a geometrical optical phenomenon. Results from wave theory are used to explain how light is refracted by a mineral. Several techniques for the measurement of optical properties of minerals use observation of light refraction.

The particle theory describes light as composed of photons of different energies with these energies related to the wavelength in the electromagnetic theory. A description of the interaction of photons with the bonding electrons in minerals can be used to explain such phenomena as refractive index, color, and pleochroism, and to interpret most spectroscopic studies.

Polarized Light

The electromagnetic theory of light is used to explain polarization phenomena. There are four types of polarized light: random, plane, 4 circular, and elliptical. Plane polarized is the most important form for the study of minerals. However, circular and elliptical forms become important in more advanced studies. Recall that the direction of energy transfer is perpendicular to the propagation for a light ray. In plane polarized light the energy transfer directions (vibration directions) are restricted to a plane parallel to and coincident with the ray path. Plane polarized light is used to study anisotropic crystals because the vibration direction of the light can be made parallel to specific directions within the crystal.

Refractive Index

The refractive index (n) is a physical property of a mineral mathematically defined as:

$$n = \frac{v_v}{v_m}$$

where,

n = Refractive index of mineral,

v_v = Velocity of light in a vacuum,

v_m = Velocity of light in the mineral.

There are three important points gained from an examination of Equation $n = \frac{v_v}{v_m}$. The refractive index of a vacuum is 1.0. Refractive index is a unitless number. Because the velocity of light cannot exceed that in a vacuum (3×10^8 m/s), the refractive index of any material is greater than 1. In routine work the velocity of light in air is substituted for the velocity of light in a vacuum in Equation $n = \frac{v_v}{v_m}$. This only amounts to a small error because the refractive index of air is approximately 1.0003.

Refractive index usually increases when matter changes from a gas to a liquid to a solid. The quantum theory of light is used to explain changes in refractive index. The refractive index is greater than one because as photons enter a mineral they are slowed by interaction with electrons. The more photons are slowed while travelling through a material, the greater the material's refractive index. Generally, any mechanism that increases electron density in a material also increases refractive index. For instance, increasing the density of a material usually increases the refractive index. However, refractive index is also closely related to bonding. In general, ionic compounds having lower values of n than covalent ones. In covalent boning more electrons are being shared by the ions than in ionic bonding; thus, more electrons are distributed through the structure and interact with the incident photons slowing them. Thus, changes in refractive index are related, albeit in a very complicated, as yet not fully understood, manner to the interaction of the incident light's photons and the material's bonding electrons.

Light Refraction

Geometrically, light travels through a crystal with the incident and transmitting rays following Snell's Law, mathematically written:

$$n_i \sin(\theta_i) = n_t \sin(\theta_t)$$

where,

n_i = Refractive index of incident media,

n_t = Refractive index of transmitting media,

θ_i = Angle of incidence,

θ_t = Angle of transmission or refraction.

Much of the science of physical optics and ray tracing is based on this formula. By analyzing Equation $n_i \sin(\theta_i) = n_t \sin(\theta_t)$ the following are noted. Any ray at normal incidence onto a material of different refractive index will be transmitted without any deviation. When both the incident and transmitting Medias have the same refractive index, the incident and transmitting angles are equal. For the general case, when the angle of incident is not 0° and the refractive indices for the two materials are unequal, the ray is refracted and the angle of refraction can be found from Equation $n_i \sin(\theta_i) = n_t \sin(\theta_t)$. There is some angle at which an incident ray is refracted at 90°; this is termed total refraction and used in refractometery to determine the refractive index of an unknown. The angle at which total refraction occurs is called the critical angle.

Every incident ray, before the critical angle is reached, has a reflected as well as a transmitted portion. For the reflected component, the angle of reflection is equal to the angle of incidence. For both rays the intensities and polarization properties can also be calculated Snell's Law is obeyed by all isotropic materials; however, it is not obeyed for random orientations in anisotropic materials.

Dispersion

The refractive index of a material depends upon the wavelength of light and can be written mathematically:

$$n = f(\gamma)$$

where,

n = Refractive index of mineral,

l = Wavelength of light.

This change in refractive index is called dispersion and is graphically. The bolder curve has a steeper slope and exhibits a greater dispersion than the lighter curve. In general, the dispersion of a liquid is greater than that of a solid. Also, in general, the higher the refractive index of a material the higher the dispersion.

For quantitative optical measurements it is necessary to find a mathematical fit to the data. Several relationships, both theoretical and empirical, have been proposed to quantify the relationship in Equation $n = f(\gamma)$. The Cauchy equation is one such equation.

$$n = c_1 + \frac{c_2}{\gamma^2} + \frac{c_3}{\gamma^4}$$

where,

n = Refractive index,

l = Wavelength of light,

C_n = Values obtained from a regression.

The rainbow, created from drops of rain, or a glass prism are familiar examples of dispersion. The refractive index of water, or glass, decreases with wavelength. White light is composed of the different wavelengths of light, and each wavelength possesses a different refractive index. Application of Snell's Law shows that a different angle of refraction occurs for different colors of light. Thus, because of dispersion and geometrical optics from Snell's Law, the colors of the rainbow are easily understood. Red light, with the lowest refractive index, is refracted the least and occurs on the outside of the rainbow. Violet light, with the highest refractive index, is refracted the most and occurs on the inside of the rainbow.

Measurement of Refractive Index

There are two general methods to measure the refractive index of transparent compounds: a direct measurement of the angle of refraction as found in Snell's Law or a comparison of the unknown material's refractive index to a known. The former method is routinely used to determine the refractive index of liquids and larger solid samples (greater than 1 mm) and on its application to anisotropic samples. While the latter method is more normally used to determine the refractive indices of smaller samples (less than 1 mm). The comparison method was developed early in this century and has become known as the immersion method. In this method a solid sample of unknown refractive index is placed into a liquid of known refractive index. By microscopic observations, explained later, the liquid is "adjusted" until its refractive index matches that of the solid, thus determining the refractive index of the unknown.

The immersion methods works on the basis of refraction of light rays as defined in Snell's Law. When the refractive index of the liquid (n_l) and solid (n_s) are the same, there is no refraction of the light rays, basically making the samples invisible (the invisible man had the ability to change his refractive index to that of air, thus becoming invisible, but how did he see?). When there is a large difference between the refractive index of the sample and liquid, light is refracted when entering and leaving the solid. The refraction of these waves makes the sample visible. The word "relief" is used to semi-quantitatively denote the difference between the refractive index of the solid and that of the liquid. Samples at or near a match are have low relief, while those farther from match have high relief. When $n_l < n_s$ the sample has positive relief; when $n_l > n_s$ the sample has negative relief.

The mineral's relief is observed in thin section mineral identification. The epoxy holding the thin section to the slide has a refractive index of approximately 1.55. Minerals in thin section with a refractive index of within +/- 0.04 will have low positive/negative relief. Minerals with a refractive index within the range of +/- 0.04 to 0.12 will have moderate positive/negative relief, while those with refractive indices above +/- 0.12 will have high positive/negative relief.

The Becke line method is the most commonly used technique to determine a match between n_l and n_s in the immersion method. A crystal is immersed in liquid between a glass slide and a coverslip. After obtaining a focused view of the crystal under the microscope in polychromatic light, the microscope is slightly de-focused by racking the stage down. Two lines, one light and one dark, appear around the grain boundaries. The light Becke line moves into the material which has the larger refractive index. Basically, the grain is functioning as a lens in the liquid. When $n_l < n_s$, the grain acts as a convex lens, refracting the light into the grain's center - the light Becke moving into the grain. When $n_l > n_s$, the grain acts as a diverging lens, sending the bright Becke line into the liquid. Another liquid is selected based upon the findings of the previous observation, and the process of bracketing the solid with the liquid continues until the Becke lines become colored, at which time another strategy is used. The Becke lines become colored when the dispersion curves for the liquid and solid intersect in the visible. The intersection points marking the precise wavelength of match were the refractive index of the liquid and solid are equal. The microscopist usually attempts to determine the refractive index at 589.3 nm, the spectra line produced by a sodium gas discharge bulb. This match is found when the Becke line that enters the grain becomes sky blue and the other orange-yellow line enters the grain. An entire progression of colors occurs, depending upon the point of intersection of the two curves.

The double-variation method is a refinement on the above technique. With it, and careful observations, refractive indices can be obtained with an accuracy and precision of +/- 0.0001. The "double" originates from the microscopist's control over both wavelength of light and temperature of the liquid. Refractive index decreases with wavelength. In this method a monochrometer is used and the precise wavelength at which a Becke line disappears can be found. Also, refractive index decreases with increasing temperature. For a material this change is mathematically written as dn/dt. The dn/dt for liquids is several orders of magnitude greater than that of solids. By raising the temperature of the liquid a few degrees centigrade, a large change occurs in its refractive index, while the crystal's refractive index does not significantly change. By making several wavelength matches at different temperature, the precision of the refractive index determination is increased. This is the chosen method, coupled with use of the spindle stage, to obtain accurate and precise optical data on single crystals.

Advanced Methods in Optical Mineralogy

Spindle stage the spindle stage is a one-axis rotation device that mounts onto the stage of a PLM. With it, any privileged vibration direction of an anisotropic crystal can be rotated into the plane of the microscope stage and brought parallel to the incident plane polarizer light, thus allowing for direct measurement of any refractive index without appreciable error due to misorientation.

When the crystal is immersed in an oil cell the mineral's refractive index can be checked against the oil's as described in the immersion method above. The main advantages of the spindle stage are: all the measurements can be made on the same crystal, the oil cell is changed and the crystal retained, thus not requiring multiple slide mounts as in the routine immersion method, and the microscopist has complete control over the mineral's orientation.

The orientation of the biaxial indicatrix can be determined in one of two ways with the spindle stage: the interference figures can be observed while the crystal is rotated about the spindle axis, or extinction data can be collected and solved to determine directly the spindle stage setting (S) and the microscope stage setting (Ms) that orients AB, OB, and ON into the plane of the microscope and parallel to the lower polarizer. The latter method has the advantage of determining 2V for the mineral to within a few tenths of a degree.

Refractometry

Refractometers are normally used to determine the single refractive index for a liquid or an isotropic solid. However, refractometers can be used to determine all the principal refractive indices for anisotropic samples. Further, all that is required is a single polished random section through the mineral. In fact, these techniques were developed and refined in the late 1800's, long before the immersion method. However, they have fallen so far out of use that most current books and research papers are filled with mistakes and lack of references to the original works.

Light is incident from the back and is reflected from the mineral / hemicylinder interface. Analysis of this reflected light and application of Snell's Law is required to determine the refractive index of the mineral. For anisotropic minerals the orientation of the plane polarized light must be controlled. Figure shows the intersection of the measurement plane (X-Y plane) of the refractometer with an anisotropic indicatrix. When the incident polarized light is made to vibrate parallel to N, its value can be found; likewise, when the light is parallel to n, its value can be found. By rotation of the sample about X and collection of N and n values, the size and shape of the optical indicatrix can be found.

The following similarity exists between the spindle stage and the refractometer: with the refractometer any vibration direction brought into the X-Y plane can be determined, and with the spindle stage the vibration direction must be brought into the plane of the microscope 28 stage. For either case, once the vibration direction is in the "measurement" plane, the next step is to orient the plane polarized light parallel to it.

Magnetic Properties

Magnetic separation is a key stage in many mineral processing plants. Understanding the mineralogy of the deposit, and how different minerals behave in a magnetic field, is important when assessing the feasibility of using magnetic separation.

Three fractions after magnetic separation: Non Magnetics, Middlings and Magnetics.

The behaviour minerals exhibit in a magnetic field is classed under one of five headings:

- Diamagnetic,
- Paramagnetic,
- Ferromagnetic,
- Anti-ferromagnetic,
- Ferrimagnetic.

These classifications define whether it is possible to separate one or more minerals from another using a magnetic separator.

Source of Magnetic Properties

The magnetic properties of any mineral reside in the electrons of the atoms or ions. In accordance with the principles of wave-mechanics, the electron, moving in a closed path about the nucleus, is considered as a current behaving like a wave, and this moving current generates a magnetic field. When a crystal is placed in an external non-uniform magnetic field, there is a force working to align the magnetic fields of the atoms to produce a magnetic moment for the whole crystal. The magnetic susceptibility χ, is the ratio of the resulting magnetic moment, M, to the strength of the external field, H.

$$\chi = M/H$$

Diamagnetism and Paramagnetism

Diamagnetic minerals have a small negative value of χ and are slightly repelled by the magnetic field. In contrast, paramagnetic minerals have a small positive value of χ and are weakly attracted by the field.

Paramagnetism is associated with the spins of the electrons, whereas diamagnetism is related to their distribution in space. Diamagnetism is a property possessed by all atoms. However, when the atom contains an odd number of electrons, or has incomplete electron shells (as in the transition elements) imbalance of the electron spins causes the paramagnetic effect to overshadow the diamagnetic part of the total magnetic susceptibility. Paramagnetism is also found in metals where there is a cloud of free conduction electrons.

Paramagnetic minerals separated using a Wet High Intensity Magnetic Separator.

This behaviour is only applied generally to crystals because the internal crystal field, as a whole modifies the magnetic effects. The electronic energy levels in a crystal are described as being split and the total magnetic susceptibility depends on the distribution of the electrons in the different levels. Therefore, in complex compounds, it is not possible to predict the magnetic properties. In minerals, iron-bearing structures are paramagnetic. However, there are paramagnetic minerals without iron.

The differences in magnetic susceptibility are sufficient to enable separation using high-intensity magnetic separators. Bismuth is the only one example of a *diamagnetic* mineral. *Paramagnetic* minerals are more widespread and include Hematite and Franklinite.

Ferromagnetism

Ferromagnetic minerals possess a magnetic moment even in the absence of an applied magnetic field. They are strongly attracted by even a weak magnetic field and remain permanently magnetised. However, ferromagnetic substances also exist in unmagnetised conditions when, at ordinary temperatures, the electronic magnetic moments are permanently in alignment as a result of interaction between neighbouring atoms. Iron, Cobalt, Nickel and Pyrrhotite are typical examples of *ferromagnetic* minerals.

Antiferromagnetism and Ferrimagnetism

The way in which electrons align in certain crystals produces either an antiferromagnetic or ferrimagnetic effect:

- Antiferromagnetism: This occurs when adjacent atoms interact in a manner to align the spins in parallel but opposed directions called *antiparallel spins*. The two sets of moments cancel one another and there is no permanent magnetic moment. Typical examples include metals such as Chromium and oxides such as Nickel Oxide (NiO).
- Ferrimagnetism: Antiparallel alignment, in which the components in opposite directions are not equal, results in a permanent moment which is *ferrimagnetism*. Magnetite is a important example of a ferrimagnetic mineral.

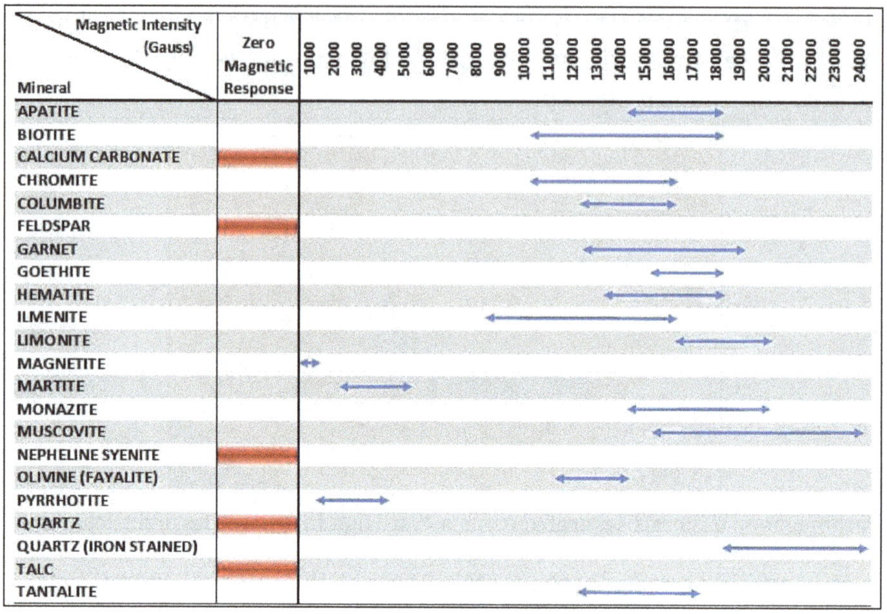

Magnetic properties of minerals.

Magnetic Separation of Minerals

Understanding the magnetic properties of the minerals in a particular deposit is vital when optimising magnetic separation. The wide range of magnetic separator designs use different strengths and types of magnetic field to produce a separation. Several designs of magnetic separator may feature in a single mineral processing operation.

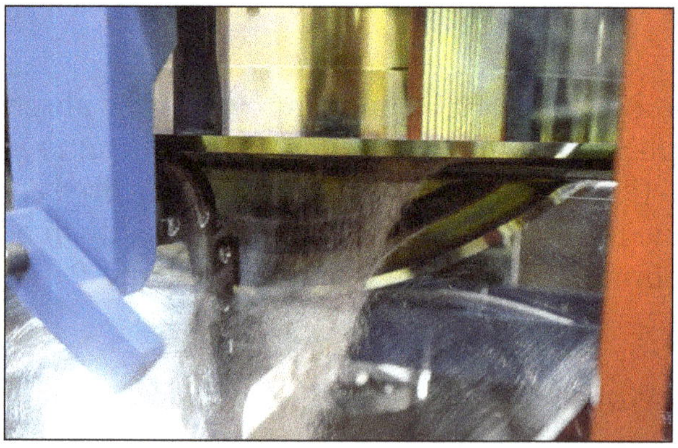
Minerals with varying magnetic properties separated using a Magnetic Disc Separator.

The most commonly used high-intensity magnetic separators in the mineral processing industry are:

- Dry Processing:
    - Rare Earth Roll Magnetic Separator,
    - Induced Roll Magnetic Separator,
    - Magnetic Disc Separator.

- Wet Processing:
  - Wet High Intensity Magnetic Separator,
  - Electromagnetic Filter.

Table: High intensity magnetic separators for mineral processing.

Magnetic Separator	Process	Particles Size Margin	Gauss Range
Electromagnetic filter	Wet	0-500 microns	0-10000
Induced roll magnetic separator (IMR)	Dry	65-1000 microns	0-22000
Magnetic Disc Serarator(MDS)	Dry	65-1000 microns	0-20000
Rare Earth roll separator (RE Roll)	Dry	65 microns-10mm	14000
Wet high intensity magnetic separator (WHIMS)	Dry	0-800 microns	0-18000

As traditional mineral reserves are depleted and more complex deposits are mined, understanding the mineralogy in terms of magnetic susceptibility is vitally important. Bunting engineers work closely with mineral processors and plant designers, often at the project feasibility stage, providing advice on the optimum design of magnetic separator and stage within the process.

Mineral Strength

Tenacity

Tenacity describes the reaction of a mineral to stress such as crushing, bending, breaking, or tearing. Certain minerals react differently to each type of stress. Since tenacity is composed of several reactions to various stresses, it is possible for a mineral to have more than one type of tenacity. The different forms of tenacity are:

- Brittle: If a mineral is hammered and the result is a powder or small crumbs, it is considered brittle. Brittle minerals leave a fine powder if scratched, which is the way to test a mineral to see if it is brittle. The majority of all minerals are brittle. An example is Quartz.(Minerals that are not brittle may be referred to as Nonbrittle minerals.)
- Sectile: Sectile minerals can be separated with a knife, much like wax but usually not as soft. An example is Gypsum.
- Malleable: If a mineral can be flattened by pounding with a hammer, it is malleable. All true metals are malleable. An example is Silver.
- Ductile: A mineral that can be stretched into a wire is ductile. All true metals are ductile. An example is Gold.

- Flexible but inelastic: Any mineral that can be bent, but remains in the new position after it is bent are flexible but inelastic. If the term flexible is singularly used, it implies flexible but inelastic. An example is Copper.
- Flexible and elastic: When flexible and elastic minerals are bent, they spring back to their original position. All fibrous minerals and some acicular minerals belong in this category. An example is Chrysotile Serpentine.

How to Test using Tenacity

When testing for brittleness, one should scratch the specimen with a harder mineral or material, and see if the mineral leaves a fine powder. If fine powder is scratched off the mineral, it is brittle. If just a furrow is left without any fine powder, the mineral is nonbrittle. Brittleness and hardness are usually tested simultaneously, since the same procedure can be conducted for both of them. When testing for brittleness, do not scratch a noticeable surface area, as this may decrease a specimen's value. Additionally, the surface being tested by scratching should not be dirty, coated, or tarnished.

If a mineral is suspected as being sectile, insert a knife into the mineral and see if it goes through. If the knife penetrates, the mineral is sectile. Sectility can also be tested with a long fingernail, by slowly pushing it into the specimen to see if it enters. If the fingernail goes through, the mineral is sectile.

To check for flexibility, slight pressure should be placed on the mineral to see if it bends. If it moves to the new position after stress is released, it is inelastic. If it goes back to its original position after stress is released, it is elastic. Even some flexible minerals will break if there is too much stress is applied. Therefore, they should be bent carefully and slowly, and only a minimum amount. Only few minerals are flexible, so thin minerals may be damaged even when slight flexing pressure is applied.

Cleavage, Fracture and Parting

Cleavage, fracture, and parting all have to do with the positioning of atoms in a mineral and how it breaks when put under stress.

Cleavage

In mineral terms, cleavage describes how a crystal breaks when subject to stress on a particular plane. If part of a crystal breaks due to stress and the broken piece retains a smooth plane or crystal shape, the mineral has cleavage. A mineral that never produces any crystallized fragments when broken off has no cleavage.

Cleavage is often measured by three factors:

- Quality of Cleavage,
- Number of Sides Exhibiting Cleavage,
- Cleavage Habit.

Quality of Cleavage

Quality of cleavage can be categorized into five qualities:

- Perfect,
- Good,
- Poor,
- Indiscernible (Indistinct),
- None.

Minerals with perfect cleavage will cleave without leaving any rough surfaces; a full, smooth plane is formed where the crystal broke. Minerals with good cleavage also leave smooth surfaces, but often leave over minor residual rough surfaces. On minerals with poor cleavage, the smooth crystal edge is not very visible, since the rough surface is dominant. If a mineral exhibits cleavage, but it so poor that it is hardly noticeable, it has "indiscernible" cleavage. Minerals with no cleavage never exhibit any cleavage, thus broken surfaces are fractured and rough. Categorization of cleavage qualities is not scientifically affirmed. The above categorization is used by most mineral references, but some guides categorize cleavage in three or four groups, and may give them different names, such as "excellent" and "distinct".

Number of Sides Exhibiting Cleavage

Many minerals exhibit cleavage only on one side and some may exhibit different quality cleavage on different crystal sides. The following criteria may be expected when analyzing the cleavage of any particular mineral:

- One Direction,
- Two Directions,
- Three Directions,
- All Directions.

These identify how many "directions", or planes, the crystal is exhibiting the cleavage on. Each direction signifies the two opposite sides of a three-dimensional figure, (since opposite sides will always exhibit the same cleavage properties). If a mineral has cleavage in three directions, then every side of the mineral has cleavage (i.e. length, width, and height). If a mineral occurs in modified crystals with more than six sides (i.e. an octahedron) and exhibits cleavage on all the sides, than it has cleavage in "all directions".

Combining the cleavage level together with the number of sides will measure the cleavage of a mineral. For example, if a mineral has Good Cleavage, Two Directions, this means that it has good cleavage on four out of six sides (while the other two sides exhibit no cleavage). If a mineral has Perfect Cleavage, One Direction; Poor Cleavage, Two Directions, this means that the mineral has perfect cleavage on two sides, and poor cleavage on the other four.

Cleavage quality is measured in numbers, then the amount of sides, separated by a comma. 1 is perfect cleavage, 2 is good cleavage, and 3 is poor cleavage. If the cleavage of a mineral is written as 1, 2 the mineral has perfect cleavage in two directions. If all sides of mineral have the same cleavage and the mineral often occurs in modified crystals with more than six sides, than All Sides is written instead of a number. If a mineral exhibits different cleavage on different crystal planes, there will be two cleavage indicators separated by a semi-colon (;). For example, if the cleavage of a mineral is written as 1, 2; 3, 1, than it has perfect cleavage in two directions, and poor cleavage in one other direction. If a mineral exhibits indistinct or no cleavage, Indiscernible or None is written in the cleavage field.

Cleavage Habit

Different habits of cleavage exist on different minerals, depending on their mode of crystallization. These forms of cleavage are:

- Basal cleavage: Cleavage exhibited on a horizontal plane of the mineral by way of its base. Minerals with basal cleavage can sometimes be “peeled”. Examples of basal cleavage are the mica minerals.
- Cubic cleavage: Cleavage exhibited on minerals of the isometric crystal system that are crystallized as cubes. In this method of cleavage, small cubes evenly break off of an existing cube. An example is Galena.
- Octahedral cleavage: Cleavage exhibited on minerals of the isometric crystal system that are crystallized as octahedrons. In this method of cleavage, flat, triangular “wedges” peel off of an existing octahedron. An example is Fluorite.
- Prismatic cleavage: Cleavage exhibited on some prismatic minerals in which a crystal cleaves as thin, vertical, prismatic crystals off of the original prism. An example is Aegirine.
- Pinicoidal cleavage: Cleavage exhibited on some prismatic and tabular minerals in which a crystal cleaves on the pinacoidal plane, which is the third dimension aside from the basal and prismatic sides. An example is Barite.
- Rhombohedral cleavage: Cleavage exhibited on minerals crystallizing in the hexagonal crystal system as rhombohedrons, in which small rhombohedrons break off of the existing rhombohedron. An example is Calcite.

Parting

Parting is characteristically similar to cleavage. It is easily confused with cleavage, and it may be present on minerals that do not exhibit any cleavage. There are two causes of parting:

- Two separate pressures pushed toward the center of a crystal after its formation, causing the crystal interior to evenly dislodge on a flat, smooth plane.
- Twinned crystals that separated from one another, leaving a flat, smooth plane.

With enough perception, a distinction can be made between parting and cleavage. If fracture marks are present on a crystal in addition to a cleaved plane, the "cleaved" surface is usually the result of parting, not cleavage. An outline of a crystal etched in a mineral is also the result of parting, in the form of twinned crystals that separated. In general, one need not worry about confusing parting with cleavage. Parting is uncommon, and it can usually be determined by the distinguishing characteristics.

Fracture

Fracture is the characteristic mark left when a mineral chips or breaks. Cleavage and fracture differ in that cleavage is the break of a crystal face where a new face (resulting in a smooth plane) is formed, whereas fracture is the "chipping" shape of a mineral. All minerals exhibit a fracture, even those that exhibit cleavage. If a mineral with cleavage is chipped a certain way, it will fracture rather than cleave.

There are several terms to describe the various mineral fractures:

- Conchoidal: Fracture resembling a semicircular shell, with a smooth, curved surface. An example of conchoidal fracture can be seen in broken glass. This fracture is also known as "shelly" in some reference guides.
- Uneven: Fracture that leaves a rough or irregular surface.
- Hackly: Fracture that resembles broken metal, with rough, jagged, points. True metals exhibit this fracture. (This fracture is also known as "jagged").
- Splintery: Fracture that forms elongated splinters. All fibrous minerals fall into this category.
- Earthy or crumbly: Fracture of minerals that crumble when broken.
- Even or smooth: Fracture that forms a smooth surface.
- Subconchoidal: Fracture that falls somewhere between conchoidal and even; smooth with irregular rounded corners.

Almost all minerals have a characteristic fracture. Some minerals of the same species may exhibit a different fracture, but this is rare.

How to use Cleavage, Parting and Fracture as an Identification Mark

A specimen need not be broken to check its fracture habit or cleavage. Rather, it should be checked for areas of stress where it could have broken or chipped. Fracture marks are rarely present on minerals with good or excellent cleavage. Minerals with poor cleavage will fracture more often than those with good or perfect cleavage.

How to Testing using Cleavage, Parting and Fracture

Observe the mineral to see if it has any cleaved surfaces or fractured edges. If it has cleaved surfaces, the quality of the smoothness on the surface should be noted. If there are no visible cleaved surfaces, it does not mean the mineral does not exhibit cleavage. It is possible that particular specimen did not cleave. On such a specimen, it is only possible to check cleavage by chipping off a piece. This should be done gently and carefully in an area which will not degrade its value. If there is a noticeable fracture on the mineral, it is a likely indication that the mineral probably has poor cleavage or none at all. Careful observation and experience should also be able to distinguish between a cleaved crystal and a crystal that parted.

Hardness

The Mohs' hardness scale was developed in 1822 by Frederich Mohs. This scale is a chart of relative hardness of the various minerals (1 - softest to 10 - hardest). Since hardness depends upon the crystallographic direction (ultimately on the strength of the bonds between atoms in a crystal), there can be variations in hardness depending upon the direction in which one measures this property. One of the most striking examples of this is kyanite, which has a hardness of 5.5 parallel to the 1 direction (c-axis), while it has a hardness of 7.0 parallel to the 100 direction (a-axis). Talc (1), the softest mineral on the Mohs scale has hardness greater than gypsum (2) in the direction that is perpendicular to the cleavage. Diamonds (10) also show a variation in hardness (the octahedral faces are harder than the cube faces).

Mohs' hardness is a measure of the relative hardness and resistance to scratching between minerals. Other hardness scales rely on the ability to create an indentation into the tested mineral (such as the Rockwell, Vickers, and Brinell hardness - these are used mainly to determine hardness in metals and metal alloys). The scratch hardness is related to the breaking of the chemical bonds in the material, creation of microfractures on the surface, or displacing atoms (in metals) of the mineral. Generally, minerals with covalent bonds are the hardest while minerals with ionic, metallic, or van der Waals bonding are much softer.

When doing the tests of the minerals it is necessary to determine which mineral was scratched. The powder can be rubbed or blown off and surface scratches can usually be felt by running the fingernail over the surface. One can also get a relative feel for the hardness difference between two minerals. For instance quartz will be able to scratch calcite with much greater ease than you can scratch calcite with fluorite. One must also use enough force to create the scratch (if you don't use enough force even diamond will not be able to scratch quartz - this is an area where practice is important). You also have to be careful to test the material that you think you are testing and not some small inclusion in the sample. This is where using a small hand lens can be very useful to determine if the test area is homogenous.

Mineral	Hardness		
Diamond	10		Zaire 1 cm. 14 carats

<table>
<tr><td>Corundum</td><td>9</td><td></td><td>Variety ruby, India 6 cm.</td></tr>
<tr><td>Topaz</td><td>8</td><td></td><td>Mursinsk, Russia, 5cm across Seaman Museum specimen.</td></tr>
<tr><td>Quartz</td><td>7</td><td></td><td>Variety amethyst, Guerro, Mexico 16 cm.</td></tr>
<tr><td>Orthoclase</td><td>6</td><td></td><td>Orthoclase (white) on quartz, Baveno, Italy Orthoclase crystal is 3 cm tall. Seaman museum specimen.</td></tr>
<tr><td>Apatite</td><td>5</td><td></td><td>Durango, Mexico. Crystal is 7.5 cm. tall. Seaman museum specimen.</td></tr>
<tr><td>Fluorite</td><td>4</td><td></td><td>Elmwood mine, Tennessee 2.5 cm. (note phantom).</td></tr>
</table>

Calcite	3	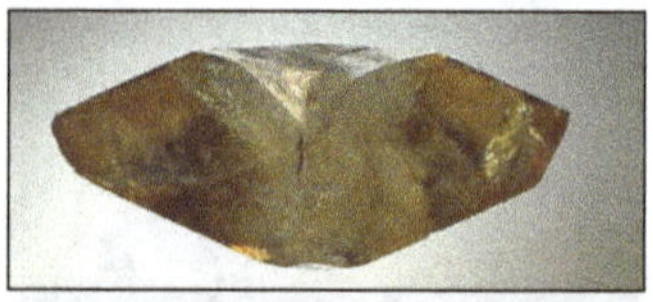	Elmwood Mine, Tennessee 8 cm. (twinned).
Gypsum	2		Wyoming 12 cm. Note "fishtail" twin on left.
Talc	1		Rope's Gold Mine, Michigan (green) 4 cm. across talc mass.

Why is Hardness Important?

The effects of high hardness are important in many fields. Abrasives are used to form and polish many substances. Diamonds are an important mineral component in cutting tools for the manufacturing of metals and other substances, forming dies for the drawing of wires, and for cutting cores in oil wells and mineral exploration. Emery - a variety of corundum, is used in many abrasive products that do not require the hardness (or expense) of diamond tools. Garnets were used as an abrasive in sandpaper. Talc is an extremely soft mineral that has been used in bath powders (talcum powder).

Mineral harness is also important in sedimentary rocks. Harder minerals tend to be able to travel longer distances down river systems. Quartz can often undergo several cycles of erosion, transportation and lithification (change of sediments to rock). Zircons are persistent minerals in the environment and can often tell geologists the types of rock that were the original source rock for metamorphic or sedimentary rocks.

Niagara Falls.

Mineral hardness can also be seen in the topography of many landscapes. Quartz bearing rocks are often more resistant to weathering and will produce the capstones that protect the tops of buttes and mesas from erosion.

Crystal Habits

The term crystal habit describes the favored growth pattern of the crystals of a mineral species, whether individually or in aggregate. It may bear little relation to the form of a single, perfect crystal of the same mineral, which would be classified according to crystal system. Subtle evidence of the crystal system to which a mineral species belongs is, however, frequently observed in the habit of the crystals which a specimen displays. The terminology used to describe crystal habit is not intended to replace the precise nomenclature of crystallography. Instead, it is intended as a supplement to this system.

Scalenohedron Calcite on Dolomite from the Elmwood Mine.

The crystals of particular minerals species sometimes form very distinctive, characteristic shapes. Crystal habit is thus often useful in identification. Although each mineral species typically forms according to a few preferred shapes, crystal habit is largely determined by the environmental conditions under which a crystal develops. For example, aqueous solutions near or surrounding a crystal contain the elemental substances which it needs to continue growth.

The direction from which a growing crystal may obtain such solutions is a factor which will affect its eventual shape. Higher environmental temperatures during formation increase ion mobility and aid in crystal formation; the rate at which the environment cools determines how much time a mineral is allowed to form large crystals. The amount of space available for a crystal to fill affects its final shape and size. Surface energy relations are also quite important to the direction of crystal growth; this process is not yet fully understood.

Adjectives used to describe the habit of individual crystals are 'equant,' 'prismatic,' and 'tabular.' Aggregates of crystals may also be termed equant or prismatic, while aggregates of thin, flat, tabular crystals may be 'bladed.' Thin sheets, flakes or scales are termed 'foliated,' 'micaceous,' and, if feathery or delicate, 'lamellar' or 'plumose.'

Crystal aggregates resembling long, slender needles, hair, or threads are termed 'acicular,' filiform,' 'capillary,' or 'fibrous.' An aggregate of crystals forming a network or lattice is 'reticulated;' one composed of branches which radiate starlike from a central point is 'stellated' while a branching and treelike mineral growth is 'dendritic.'

Colloform' crystal habits termed 'botryoidal,' 'mamillary,' and 'reniform' display spherical, bulbous or globular lumps. Smaller spherical forms are of 'pisolitic' or 'oolitic' habit; ovoid clusters or formations are 'amygdaloidal.' Tapered, column-like formations are 'stalactitic' or 'columnar' while concentrically banded formations are of 'concretionary' habit. Minerals whose flat crystal faces are covered with shallow, parallel grooves are 'striated;' a fine furry layer of crystals growing over a massive lump constitutes a formation of 'drusy' habit.

Following is a list of descriptive terms which are applied when discussing crystal habit:

Equant

A crystal which is equant or equidimensional possesses approximately the same side length in every direction. Crystals of garnet are often of equant habit.

Prismatic

A prismatic crystal is elongated in one direction like a prism. The mineral tourmaline often forms crystals of such habit.

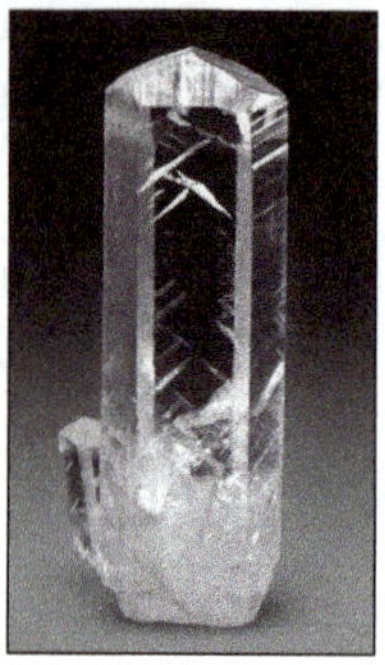

Prismatic Calcite.

Tabular

Tabular crystals appear tabular or platelike in shape.

Tabular vanadanite.

Bladed

A specimen displaying bladed habit possesses a collection of elongated, flat crystals suggestive of knife blades. Gypsum often displays crystals of bladed habit.

Bladed calcite

Micaceous

Minerals of micaceous habit form as thin, flat sheets or flakes which are easily peeled or split off the larger mass. Muscovite provides an example of micaceous habit. Crystals of foliated habit are separable into leafy structures or display leaflike projections.

Micaceous Muscovite.

Hopper

Hopper halite.

A hopper crystal is a form of crystal, defined by its "hoppered" shape. The edges of hoppered crystals are fully developed, but the interior spaces are not filled in. This results in what appears to be a hollowed out step lattice formation, as if someone had removed interior sections of the individual crystals. In fact, the "removed" sections never filled in, because the crystal was growing so rapidly that there was not enough time (or material) to fill in the gaps.

Plumose

A mineral specimen of plumose habit displays fine, feathery scales resembling plumes. 'Plumose' is derived from the Latin term pluma, or 'feather.'

Plumose Aurichalcite.

Acicular

The adjective 'acicular' means needlelike in shape. An acicular aggregate of crystals contains many long, slender crystals which may radiate out like needles or bristles from a common base. Acicular crystals are typically long and narrow like a pine leaf and seem to possess a sharp point. The mineral natrolite often exhibits acicular crystals.

Acicular agardite-Ce.

Filiform

A mineral possessing crystals of filliform habit exhibits many hairlike or threadlike filaments. "Filiform' is derived from the Latin word filum, 'thread.'

Filiform Malachite.

Fibrous

Specimens possessing fibrous habit exhibit clumps of sinewy, stringy, or hairlike fibers.

Fibrous Scolecite.

Reticulated

A mineral specimen of reticulated habit seems to display a lattice, net, or network of small crystals. The word 'reticulated' is derived from the Latin term rete, or 'net.'

Cerussite (Reticulated Habit)

Stellated

A mineral of stellated habit possesses several branches which radiate outwards from the center in a pattern resembling a star. The word 'stellated' stems from the Latin term stella, or 'star.'

Stellated Wavellite.

Dendritic

Dendritic crystals form a divergent branching structure reminiscent of an arborescent, organic growth such as a tree or a dendrite. Native copper sometimes exhibits this habit.

Dendritic Pyrolusite.

Colloform

Specimens of colloform habit exhibit spherical, rounded, or bulbous shapes. Botryoidal, reniform, and mammillary habits are subsets of this category.

Botryoidal

The word 'botryoidal' means 'resembling a bunch of grapes,' or globular. Specimens of malachite frequently provide examples of botryoidal crystals. The Greek word botrus, 'bunch of grapes,' provides the linguistic root of botryoidal.

Botryoidal Agate.

Mammillary

Samples possessing mammillary crystal habit display soft, rounded curves.

Mammillary Erythrite.

Reniform

Reniform crystal habit displays the shape of a kidney. The mineral species hematite provides samples which exemplify both mammillary and reniform habit. 'Reniform' is derived from the Latin renes, 'kidney.'

Reniform Hematite.

Oolitic

Crystals of oolitic habit form small spheres or grains which resemble fish roe. Oolites are often found in limestones.

Oolitic Calcite.

Pisolitic

A mineral of pisolitic habit develops round, pea-shaped forms. These are larger and slightly more uneven than an oolite and are usually composed of calcium carbonate. The word 'pisolitic' is derived from the Greek term pisos, 'pea.'

Pisolitic bauxite.

Stalactitic

Stalactitic or columnar crystal habit refers to the tall, tapered, columlike appearance of an icicle or a limestone stalactite. Such formations are built up by the dripping of mineral-laden solution. The minerals calcite and aragonite ($CaCO_3$) typically form stalactites. The term is derived from the Greek word stalaktos, 'dripping.'

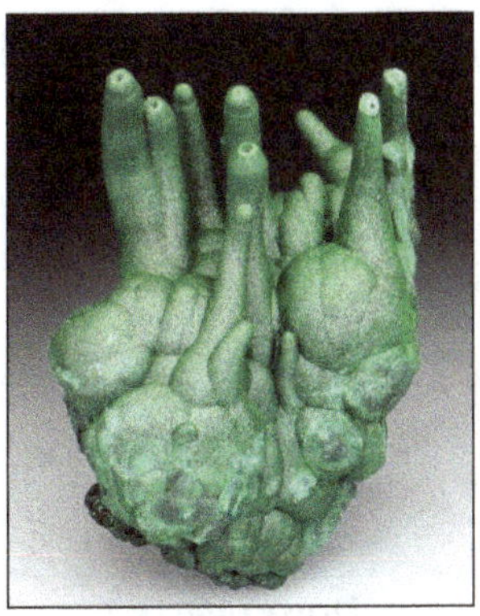

Stalactitic Malachite.

Concretionary

A concretion develops when mineral matter is concentrically deposited around a nucleus and colored and banded layers are build up. Malachite often exhibits such formations.

Pyrite Concretions.

Striated

Minerals whose crystals are of striated habit display shallow parallel grooves or lines along flat crystal faces. Pyrite often demonstrates square, striated crystals.

Striated pyrite.

Drusy

A sample exhibiting drusy habit displays a surface covered with a fine furry layer of tiny crystals.

Drusy quartz.

Massive

Massive or earthy habit describes a large, lumpy mass which has no apparent crystal form. In such a sample the crystals are too tiny to be observable by the eye and are interlocked and mingled; the specimen lacks visible crystals.

Massive Amazonite.

Electrical Properties

Three electrical properties are applicable to minerals: Conduction, Pyroelectricity, and Piezoelectricity:

- Conduction: Conduction in mineral terms is defined as the ability of a mineral to conduct electricity. Only a very small number of minerals are good conductors; they are the metallic elements and the mineral Graphite. These conductors can be placed between a wire carrying electricity, and the electricity will pass through. Conduction is an important property that can distinguish true metals from metallic looking sulfides and oxides.
- Pyroelectricity: Pyroelectricity describes the ability of a mineral to develop electrical charges when exposed to temperature changes. Some minerals develop an electrical charge when heated, others when cooled.
- Piezoelectricity: Piezoelectricity describes the ability of a mineral to develop electrical charges when put under stress. Piezoelectric minerals will develop charges when rubbed or struck repeatedly. Conduction is very useful in distinguishing true metals, but pyroelectricity and piezoelectricity are not practical testing methods for normal mineral collectors.

References

- Mineral-physics: geo-prose.com, Retrieved 07, July 2020
- Magnetic-properties-of-minerals: mastermagnets.com, Retrieved 11, February 2020
- Tenacity, property, resource: minerals.net, Retrieved 14, August 2020
- Cleavage-fracture-parting, property, resource: minerals.net, Retrieved 03, June 2020
- Collectors-corner: minsocam.org, Retrieved 15, July 2020
- Crystal-habits-and-forms: geologyin.com, Retrieved 02, January 2020
- ElectricalProperties, property, resource: minerals.net, Retrieved 07, March 2020

Chapter 4

Classification of Minerals

Minerals are classified according to their chemical properties, with the anion being the chemical basis for classification. Silicate minerals, native element minerals, oxide minerals, halide minerals, sulfate minerals and phosphate minerals are the classes of minerals that are covered in this chapter in a clear and concise manner.

Mineral

A mineral is a naturally occurring chemical compound. Most often, they are crystalline and abiogenic in origin. A mineral is different from a rock, which can be an aggregate of minerals or non-minerals and does not have one specific chemical composition, as a mineral does. The exact definition of a mineral is under debate, especially with respect to the requirement that a valid species be abiogenic, and to a lesser extent with regard to it having an ordered atomic structure.

Amethyst, a variety of quartz.

The study of minerals is called mineralogy. There are over 5,300 known mineral species; over 5,070 of these have been approved by the International Mineralogical Association (IMA). The silicate minerals compose over 90% of the Earth's crust. The diversity and abundance of mineral species is controlled by the Earth's chemistry. Silicon and oxygen constitute approximately 75% of the Earth's crust, which translates directly into the predominance of silicate minerals.

Minerals are distinguished by various chemical and physical properties. Differences in chemical composition and crystal structure distinguish the various species, which were determined by the mineral's geological environment when formed. Changes in the temperature, pressure, or bulk composition of a rock mass cause changes in its minerals.

Minerals can be described by their various physical properties, which are related to their chemical structure and composition. Common distinguishing characteristics include crystal structure and habit, hardness, lustre, diaphaneity, colour, streak, tenacity, cleavage, fracture, parting, and specific gravity. More specific tests for describing minerals include magnetism, taste or smell, radioactivity and reaction to acid.

Minerals are classified by key chemical constituents; the two dominant systems are the Dana classification and the Strunz classification. The silicate class of minerals is subdivided into six subclasses by the degree of polymerization in the chemical structure. All silicate minerals have a base unit of a $[SiO_4]^{4-}$ silica tetrahedron—that is, a silicon cation coordinated by four oxygen anions, which gives the shape of a tetrahedron. These tetrahedra can be polymerized to give the subclasses: orthosilicates (no polymerization, thus single tetrahedra), disilicates (two tetrahedra bonded together), cyclosilicates (rings of tetrahedra), inosilicates (chains of tetrahedra), phyllosilicates (sheets of tetrahedra), and tectosilicates (three-dimensional network of tetrahedra). Other important mineral groups include the native elements, sulfides, oxides, halides, carbonates, sulfates, and phosphates.

Definition

Basic Definition

The general definition of a mineral encompasses the following criteria:

- Naturally occurring.
- Stable at room temperature.
- Represented by a chemical formula.
- Usually abiogenic (not resulting from the activity of living organisms).
- Ordered atomic arrangement.

The first three general characteristics are less debated than the last two. The first criterion means that a mineral has to form by a natural process, which excludes anthropogenic compounds. Stability at room temperature, in the simplest sense, is synonymous to the mineral being solid. More specifically, a compound has to be stable or metastable at 25 °C. Classical examples of exceptions to this rule include native mercury, which crystallizes at −39 °C, and water ice, which is solid only below 0 °C; as these two minerals were described prior to 1959, they were grandfathered by the International Mineralogical Association (IMA). Modern advances have included extensive study of liquid crystals, which also extensively involve mineralogy. Minerals are chemical compounds, and as such they can be described by fixed or a variable formula. Many mineral groups and species are composed of a solid solution; pure substances are not usually found because of contamination or chemical substitution. For example, the olivine group is described by the variable formula $(Mg, Fe)_2SiO_4$, which is a solid solution of two end-member species, magnesium-rich forsterite and iron-rich fayalite, which are described by a fixed chemical formula. Mineral species themselves could have a variable compositions, such as the sulfide mackinawite, $(Fe, Ni)_9S_8$, which is mostly a ferrous sulfide, but has a very significant nickel impurity that is reflected in its formula.

The requirement that a valid mineral species be abiogenic has also been described as similar to being inorganic; however, this criterion is imprecise and organic compounds have been assigned a separate classification branch. Finally, the requirement of an ordered atomic arrangement is usually synonymous with crystallinity; however, crystals are also periodic, so the broader criterion is used instead. An ordered atomic arrangement gives rise to a variety of macroscopic physical properties, such as crystal form, hardness, and cleavage. There have been several recent proposals to amend the definition to consider biogenic or amorphous substances as minerals. The formal definition of a mineral approved by the IMA in 1995:

"A mineral is an element or chemical compound that is normally crystalline and that has been formed as a result of geological processes."

In addition, biogenic substances were explicitly excluded:

"Biogenic substances are chemical compounds produced entirely by biological processes without a geological component (e.g., urinary calculi, oxalate crystals in plant tissues, shells of marine molluscs, etc.) and are not regarded as minerals. However, if geological processes were involved in the genesis of the compound, then the product can be accepted as a mineral."

Recent Advances

Mineral classification schemes and their definitions are evolving to match recent advances in mineral science. Recent changes have included the addition of an organic class, in both the new Dana and the Strunz classification schemes. The organic class includes a very rare group of minerals with hydrocarbons. The IMA Commission on New Minerals and Mineral Names adopted in 2009 a hierarchical scheme for the naming and classification of mineral groups and group names and established seven commissions and four working groups to review and classify minerals into an official listing of their published names. According to these new rules, "mineral species can be grouped in a number of different ways, on the basis of chemistry, crystal structure, occurrence, association, genetic history, or resource, for example, depending on the purpose to be served by the classification."

The Nickel (1995) exclusion of biogenic substances was not universally adhered to. For example, Lowenstam (1981) stated that "organisms are capable of forming a diverse array of minerals, some of which cannot be formed inorganically in the biosphere." The distinction is a matter of classification and less to do with the constituents of the minerals themselves. Skinner (2005) views all solids as potential minerals and includes biominerals in the mineral kingdom, which are those that are created by the metabolic activities of organisms. Skinner expanded the previous definition of a mineral to classify "element or compound, amorphous or crystalline, formed through *biogeochemical* processes," as a mineral.

Recent advances in high-resolution genetics and X-ray absorption spectroscopy are providing revelations on the biogeochemical relations between microorganisms and minerals that may make Nickel's (1995) biogenic mineral exclusion obsolete and Skinner's (2005) biogenic mineral inclusion a necessity. For example, the IMA commissioned "Environmental Mineralogy and Geochemistry Working Group" deals with minerals in the hydrosphere, atmosphere, and biosphere. The group's scope includes mineral-forming microorganisms, which exist on nearly every rock, soil, and particle surface spanning the globe to depths of at least 1600 metres below the sea floor and

70 kilometres into the stratosphere (possibly entering the mesosphere). Biogeochemical cycles have contributed to the formation of minerals for billions of years. Microorganisms can precipitate metals from solution, contributing to the formation of ore deposits. They can also catalyze the dissolution of minerals.

Prior to the International Mineralogical Association's listing, over 60 biominerals had been discovered, named, and published. These minerals (a sub-set tabulated in Lowenstam (1981)) are considered minerals proper according to the Skinner (2005) definition. These biominerals are not listed in the International Mineral Association official list of mineral names, however, many of these biomineral representatives are distributed amongst the 78 mineral classes listed in the Dana classification scheme. Another rare class of minerals (primarily biological in origin) include the mineral liquid crystals that have properties of both liquids and crystals. To date over 80,000 liquid crystalline compounds have been identified.

The Skinner (2005) definition of a mineral takes this matter into account by stating that a mineral can be crystalline or amorphous, the latter group including liquid crystals. Although biominerals and liquid mineral crystals, are not the most common form of minerals, they help to define the limits of what constitutes a mineral proper. The formal Nickel (1995) definition explicitly mentioned crystallinity as a key to defining a substance as a mineral. A 2011 article defined icosahedrite, an aluminium-iron-copper alloy as mineral; named for its unique natural icosahedral symmetry, it is a quasicrystal. Unlike a true crystal, quasicrystals are ordered but not periodic.

Rocks, Ores, and Gems

Schist is a metamorphic rock characterized by an abundance of platy minerals. In this example, the rock has prominent sillimanite porphyroblasts as large as 3 cm (1.2 in).

Minerals are not equivalent to rocks. A rock is either an aggregate of one or more minerals, or not composed of minerals at all. Rocks like limestone or quartzite are composed primarily of one mineral—calcite or aragonite in the case of limestone, and quartz in the latter case. Other rocks can be defined by relative abundances of key (essential) minerals; a granite is defined by proportions of quartz, alkali feldspar, and plagioclase feldspar. The other minerals in the rock are termed accessory, and do not greatly affect the bulk composition of the rock. Rocks can also be composed entirely of non-mineral material; coal is a sedimentary rock composed primarily of organically derived carbon.

In rocks, some mineral species and groups are much more abundant than others; these are termed the rock-forming minerals. The major examples of these are quartz, the feldspars, the micas, the amphiboles, the pyroxenes, the olivines, and calcite; except the last one, all of the minerals are silicates. Overall, around 150 minerals are considered particularly important, whether in terms of their abundance or aesthetic value in terms of collecting.

Commercially valuable minerals and rocks are referred to as industrial minerals. For example, muscovite, a white mica, can be used for windows (sometimes referred to as isinglass), as a filler, or as an insulator. Ores are minerals that have a high concentration of a certain element, typically a metal. Examples are cinnabar (HgS), an ore of mercury, sphalerite (ZnS), an ore of zinc, or cassiterite (SnO_2), an ore of tin. Gems are minerals with an ornamental value, and are distinguished from non-gems by their beauty, durability, and usually, rarity. There are about 20 mineral species that qualify as gem minerals, which constitute about 35 of the most common gemstones. Gem minerals are often present in several varieties, and so one mineral can account for several different gemstones; for example, ruby and sapphire are both corundum, Al_2O_3.

Nomenclature and Classification

Minerals are classified by variety, species, series and group, in order of increasing generality. The basic level of definition is that of mineral species, which is distinguished from other species by specific and unique chemical and physical properties. For example, quartz is defined by its formula, SiO_2, and a specific crystalline structure that distinguishes it from other minerals with the same chemical formula (termed polymorphs). When there exists a range of composition between two minerals species, a mineral series is defined. For example, the biotite series is represented by variable amounts of the endmembers phlogopite, siderophyllite, annite, and eastonite. In contrast, a mineral group is a grouping of mineral species with some common chemical properties that share a ccrystal structure. The pyroxene group has a common formula of $XY(Si,Al)_2O_6$, where X and Y are both cations, with X typically bigger than Y; the pyroxenes are single-chain silicates that crystallize in either the orthorhombic or monoclinic crystal systems. Finally, a mineral variety is a specific type of mineral species that differs by some physical characteristic, such as colour or crystal habit. An example is amethyst, which is a purple variety of quartz.

Two common classifications, Dana and Strunz, are used for minerals; both rely on composition, specifically with regards to important chemical groups, and structure. James Dwight Dana, a leading geologist of his time, first published his *System of Mineralogy* in 1837; as of 1997, it is in its eighth edition. The Dana classification assigns a four-part number to a mineral species. Its class number is based on important compositional groups; the type gives the ratio of cations to anions in the mineral; and the last two numbers group minerals by structural similarity within a given type or class. The less commonly used Strunz classification, named for German mineralogist Karl Hugo Strunz, is based on the Dana system, but combines both chemical and structural criteria, the latter with regards to distribution of chemical bonds.

As of January 2016, 5,090 mineral species are approved by the IMA. They are most commonly named after a person (45%), followed by discovery location (23%); names based on chemical composition (14%) and physical properties (8%) are the two other major groups of mineral name etymologies.

Fake Minerals

The fake minerals are minerals (or gems, that is to say outstanding minerals) not natural, man-made. This can be a natural mineral transformed by man into another, or an entirely artificial mineral. Also referred to as false to the synonyms. In short, counterfeiting may be partial (sample processed) or total (sample created by humans) or cover the name given to the sample. They have always existed, their marketing is growing very rapidly. There are now numerous examples of fakes in mineralogy and gemology. If there are criteria for authentication of minerals, it is sometimes difficult to distinguish between fake and genuine ones. Fake samples are sold for real deception, but when the infringement is announced or found, it may have a financial interest, decorative or teaching for the buyer.

Mineral Chemistry

Hübnerite, the manganese-rich end-member of the wolframite series, with minor quartz in the background.

The abundance and diversity of minerals is controlled directly by their chemistry, in turn dependent on elemental abundances in the Earth. The majority of minerals observed are derived from the Earth's crust. Eight elements account for most of the key components of minerals, due to their abundance in the crust. These eight elements, summing to over 98% of the crust by weight, are, in order of decreasing abundance: oxygen, silicon, aluminium, iron, magnesium, calcium, sodium and potassium. Oxygen and silicon are by far the two most important — oxygen composes 46.6% of the crust by weight, and silicon accounts for 27.7%.

The minerals that form are directly controlled by the bulk chemistry of the parent body. For example, a magma rich in iron and magnesium will form mafic minerals, such as olivine and the pyroxenes; in contrast, a more silica-rich magma will crystallize to form minerals that incorporate more SiO_2, such as the feldspars and quartz. In a limestone, calcite or aragonite (both $CaCO_3$) form because the rock is rich in calcium and carbonate. A corollary is that a mineral will not be found in a rock whose bulk chemistry does not resemble the bulk chemistry of a given mineral with the exception of trace minerals. For example, kyanite, Al_2SiO_5 forms from the metamorphism of aluminium-rich shales; it would not likely occur in aluminium-poor rock, such quartzite.

The chemical composition may vary between end member species of a solid solution series. For example, the plagioclase feldspars comprise a continuous series from sodium-rich end member albite ($NaAlSi_3O_8$) to calcium-rich anorthite ($CaAl_2Si_2O_8$) with four recognized intermediate varieties between them (given in order from sodium- to calcium-rich): oligoclase, andesine, labradorite, and bytownite. Other examples of series include the olivine series of magnesium-rich forsterite and iron-rich fayalite, and the wolframite series of manganese-rich hübnerite and iron-rich ferberite.

Chemical substitution and coordination polyhedra explain this common feature of minerals. In nature, minerals are not pure substances, and are contaminated by whatever other elements are present in the given chemical system. As a result, it is possible for one element to be substituted for another. Chemical substitution will occur between ions of a similar size and charge; for example, K^+ will not substitute for Si^{4+} because of chemical and structural incompatibilities caused by a big difference in size and charge. A common example of chemical substitution is that of Si^{4+} by Al^{3+}, which are close in charge, size, and abundance in the crust. In the example of plagioclase, there are three cases of substitution. Feldspars are all framework silicates, which have a silicon-oxygen ratio of 2:1, and the space for other elements is given by the substitution of Si^{4+} by Al^{3+} to give a base unit of $[AlSi_3O_8]^-$; without the substitution, the formula would be charge-balanced as SiO_2, giving quartz. The significance of this structural property will be explained further by coordination polyhedra. The second substitution occurs between Na^+ and Ca^{2+}; however, the difference in charge has to accounted for by making a second substitution of Si^{4+} by Al^{3+}.

Coordination polyhedra are geometric representation of how a cation is surrounded by an anion. In mineralogy, due its abundance in the crust, coordination polyhedra are usually considered in terms of oxygen. The base unit of silicate minerals is the silica tetrahedron — one Si^{4+} surrounded by four O^{2-}. An alternate way of describing the coordination of the silicate is by a number: in the case of the silica tetrahedron, the silicon is said to have a coordination number of 4. Various cations have a specific range of possible coordination numbers; for silicon, it is almost always 4, except for very high-pressure minerals where compound is compressed such that silicon is in six-fold (octahedral) coordination by oxygen. Bigger cations have a bigger coordination number because of the increase in relative size as compared to oxygen (the last orbital subshell of heavier atoms is different too). Changes in coordination numbers between leads to physical and mineralogical differences; for example, at high pressure such as in the mantle, many minerals, especially silicates such as olivine and garnet will change to a perovskite structure, where silicon is in octahedral coordination. Another example are the aluminosilicates kyanite, andalusite, and sillimanite (polymorphs, as they share the formula Al_2SiO_5), which differ by the coordination number of the Al^{3+}; these minerals transition from one another as a response to changes in pressure and temperature. In the case of silicate materials, the substitution of Si^{4+} by Al^{3+} allows for a variety of minerals because of the need to balance charges.

When minerals react, the products will sometimes assume the shape of the reagent; the product mineral is termed to be a pseudomorph of (or after) the reagent. Illustrated here is a pseudomorph of kaolinite after orthoclase. Here, the pseudomorph preserved the Carlsbad twinning common in orthoclase.

Changes in temperature and pressure, and composition alter the mineralogy of a rock sample. Changes in composition can be caused by processes such as weathering or metasomatism (hydrothermal alteration). Changes in temperature and pressure occur when the host rock undergoes tectonic or magmatic movement into differing physical regimes. Changes in thermodynamic conditions make it favourable for mineral assemblages to react with each other to produce new minerals; as such, it is possible for two rocks to have an identical or a very similar bulk rock chemistry without having a similar mineralogy. This process of mineralogical alteration is related to the rock cycle. An example of a series of mineral reactions is illustrated as follows.

Orthoclase feldspar ($KAlSi_3O_8$) is a mineral commonly found in granite, a plutonic igneous rock. When exposed to weathering, it reacts to form kaolinite ($Al_2Si_2O_5(OH)_4$, a sedimentary mineral, and silicic acid):

$$2\ KAlSi_3O_8 + 5\ H_2O + 2\ H^+ \rightarrow Al_2Si_2O_5(OH)_4 + 4\ H_2SiO_3 + 2\ K^+$$

Under low-grade metamorphic conditions, kaolinite reacts with quartz to form pyrophyllite ($Al_2Si_4O_{10}(OH)_2$):

$$Al_2Si_2O_5(OH)_4 + SiO_2 \rightarrow Al_2Si_4O_{10}(OH)_2 + H_2O$$

As metamorphic grade increases, the pyrophyllite reacts to form kyanite and quartz:

$$Al_2Si_4O_{10}(OH)_2 \rightarrow Al_2SiO_5 + 3\ SiO_2 + H_2O$$

Alternatively, a mineral may change its crystal structure as a consequence of changes in temperature and pressure without reacting. For example, quartz will change into a variety of its SiO_2 polymorphs, such as tridymite and cristobalite at high temperatures, and coesite at high pressures.

Physical Properties of Minerals

Classifying minerals ranges from simple to difficult. A mineral can be identified by several physical properties, some of them being sufficient for full identification without equivocation. In other cases, minerals can only be classified by more complex optical, chemical or X-ray diffraction analysis; these methods, however, can be costly and time-consuming. Physical properties applied for classification include crystal structure and habit, hardness, lustre, diaphaneity, colour, streak, cleavage and fracture, and specific gravity. Other less general tests include fluorescence, phosphorescence, magnetism, radioactivity, tenacity (response to mechanical induced changes of shape or form), piezoelectricity and reactivity to dilute acids.

Crystal Structure and Habit

Crystal structure results from the orderly geometric spatial arrangement of atoms in the internal structure of a mineral. This crystal structure is based on regular internal atomic or ionic arrangement that is often expressed in the geometric form that the crystal takes. Even when the mineral grains are too small to see or are irregularly shaped, the underlying crystal structure is always periodic and can be determined by X-ray diffraction. Minerals are typically described by their symmetry content. Crystals are restricted to 32 point groups, which differ by their symmetry. These groups are classified in turn into more broad categories, the most encompassing of these being the six crystal families.

Topaz has a characteristic orthorhombic elongated crystal shape.

These families can be described by the relative lengths of the three crystallographic axes, and the angles between them; these relationships correspond to the symmetry operations that define the narrower point groups. They are summarized below; a, b, and c represent the axes, and α, β, γ represent the angle opposite the respective crystallographic axis (e.g. α is the angle opposite the a-axis, viz. the angle between the b and c axes):

Crystal family	Lengths	Angles	Common examples
Isometric	a=b=c	$\alpha=\beta=\gamma=90°$	Garnet, halite, pyrite
Tetragonal	a=b≠c	$\alpha=\beta=\gamma=90°$	Rutile, zircon, andalusite
Orthorhombic	a≠b≠c	$\alpha=\beta=\gamma=90°$	Olivine, aragonite, orthopyroxenes
Hexagonal	a=b≠c	$\alpha=\beta=90°$, $\gamma=120°$	Quartz, calcite, tourmaline
Monoclinic	a≠b≠c	$\alpha=\gamma=90°$, $\beta\neq90°$	Clinopyroxenes, orthoclase, gypsum
Triclinic	a≠b≠c	$\alpha\neq\beta\neq\gamma\neq90°$	Anorthite, albite, kyanite

The hexagonal crystal family is also split into two crystal *systems* — the trigonal, which has a three-fold axis of symmetry, and the hexagonal, which has a six-fold axis of symmetry.

Chemistry and crystal structure together define a mineral. With a restriction to 32 point groups, minerals of different chemistry may have identical crystal structure. For example, halite (NaCl), galena (PbS), and periclase (MgO) all belong to the hexaoctahedral point group (isometric family), as they have a similar stoichiometry between their different constituent elements. In contrast, polymorphs are groupings of minerals that share a chemical formula but have a different structure. For example, pyrite and marcasite, both iron sulfides, have the formula FeS_2; however, the former is isometric while the latter is orthorhombic. This polymorphism extends to other sulfides with the generic AX_2 formula; these two groups are collectively known as the pyrite and marcasite groups.

Polymorphism can extend beyond pure symmetry content. The aluminosilicates are a group of three minerals — kyanite, andalusite, and sillimanite — which share the chemical formula Al_2SiO_5. Kyanite is triclinic, while andalusite and sillimanite are both orthorhombic and belong to the dipyramidal point group. These difference arise correspond to how aluminium is coordinated within the crystal structure. In all minerals, one aluminium ion is always in six-fold coordination by oxygen; the silicon, as a general rule is in four-fold coordination in all minerals; an exception is a case like stishovite (SiO_2, an ultra-high pressure quartz polymorph with rutile structure). In kyanite, the second aluminium is in six-fold coordination; its chemical formula can be expressed as AlAl-SiO_5, to reflect its crystal structure. Andalusite has the second aluminium in five-fold coordination ($AlAlSiO_5$) and sillimanite has it in four-fold coordination ($AlAlSiO_5$).

Differences in crystal structure and chemistry greatly influence other physical properties of the mineral. The carbon allotropes diamond and graphite have vastly different properties; diamond is the hardest natural substance, has an adamantine lustre, and belongs to the isometric crystal family, whereas as graphite is very soft, has a greasy lustre, and crystallises in the hexagonal family. This difference is accounted by differences in bonding. In diamond, the carbons are in sp^3 hybrid orbitals, which means they form a framework where each carbon is covalently bonded to four neighbours in a tetrahedral fashion; on the other hand, graphite is composed of sheets of carbons in sp^2 hybrid orbitals, where each carbon is bonded covalently to only three others. These sheets are held together by much weaker van der Waals forces, and this discrepancy translates to big macroscopic differences.

Contact twins, as seen in spinel.

Twinning is the intergrowth of two or more crystal of a single mineral species. The geometry of the twinning is controlled by the mineral's symmetry. As a result, there are several types of twins, including contact twins, reticulated twins, geniculated twins, penetration twins, cyclic twins, and polysynthetic twins. Contact, or simple twins, consist of two crystals joined at a plane; this type of twinning is common in spinel. Reticulated twins, common in rutile, are interlocking crystals resembling netting. Geniculated twins have a bend in the middle that is caused by start of the twin. Penetration twins consist of two single crystals that have grown into each other; examples of this twinning include cross-shaped staurolite twins and Carlsbad twinning in orthoclase. Cyclic twins are caused by repeated twinning around a rotation axis. It occurs around three, four, five, six, or eight-fold axes, and the corresponding patterns are called threelings, fourlings, fivelings, sixlings, and eightlings. Sixlings are common in aragonite. Polysynthetic twins are similar to cyclic twinning by the presence of repetitive twinning; however, instead of occurring around a rotational axis, it occurs along parallel planes, usually on a microscopic scale.

Crystal habit refers to the overall shape of crystal. Several terms are used to describe this property. Common habits include acicular, which described needlelike crystals like in natrolite, bladed, dendritic (tree-pattern, common in native copper), equant, which is typical of garnet, prismatic (elongated in one direction), and tabular, which differs from bladed habit in that the former is platy whereas the latter has a defined elongation. Related to crystal form, the quality of crystal faces is diagnostic of some minerals, especially with a petrographic microscope. Euhedral crystals have a defined external shape, while anhedral crystals do not; those intermediate forms are termed subhedral.

Hardness

The hardness of a mineral defines how much it can resist scratching. This physical property is controlled by the chemical composition and crystalline structure of a mineral. A mineral's hardness is not necessarily constant for all sides, which is a function of its structure; crystallographic weakness renders some directions softer than others. An example of this property exists in kyanite, which has a Mohs hardness of 5½ parallel to but 7 parallel to.

The most common scale of measurement is the ordinal Mohs hardness scale. Defined by ten indicators, a mineral with a higher index scratches those below it. The scale ranges from talc, a phyllosilicate, to diamond, a carbon polymorph that is the hardest natural material. The scale is provided below:

Mohs hardness	Mineral	Chemical formula
1	Talc	$Mg_3Si_4O_{10}(OH)_2$
2	Gypsum	$CaSO_4{\cdot}2H_2O$
3	Calcite	$CaCO_3$
4	Fluorite	CaF_2
5	Apatite	$Ca_5(PO_4)_3(OH,Cl,F)$
6	Orthoclase	$KAlSi_3O_8$
7	Quartz	SiO_2
8	Topaz	$Al_2SiO_4(OH,F)_2$
9	Corundum	Al_2O_3
10	Diamond	C

Lustre and Diaphaneity

Pyrite has a metallic lustre.

Lustre indicates how light reflects from the mineral's surface, with regards to its quality and intensity. There are numerous qualitative terms used to describe this property, which are split into metallic and non-metallic categories. Metallic and sub-metallic minerals have high reflectivity like metal; examples of minerals with this lustre are galena and pyrite. Non-metallic lustres include: adamantine, such as in diamond; vitreous, which is a glassy lustre very common in silicate minerals; pearly, such as in talc and apophyllite, resinous, such as members of the garnet group, silky which common in fibrous minerals such as asbestiform chrysotile.

The diaphaneity of a mineral describes the ability of light to pass through it. Transparent minerals do not diminish the intensity of light passing through it. An example of such a mineral is muscovite (potassium mica); some varieties are sufficiently clear to have been used for windows. Translucent minerals allow some light to pass, but less than those that are transparent. Jadeite and nephrite (mineral forms of jade are examples of minerals with this property). Minerals that do not allow light to pass are called opaque.

The diaphaneity of a mineral depends on thickness of the sample. When a mineral is sufficiently thin (e.g., in a thin section for petrography), it may become transparent even if that property is not seen in hand sample. In contrast, some minerals, such as hematite or pyrite are opaque even in thin-section.

Colour and Streak

Colour is typically not a diagnostic property of minerals. Shown are green uvarovite (left) and red-pink grossular (right), both garnets. The diagnostic features would include dodecahedral crystals, resinous lustre, and hardness around 7.

Colour is the most obvious property of a mineral, but it is often non-diagnostic. It is caused by electromagnetic radiation interacting with electrons (except in the case of incandescence, which does not apply to minerals). Two broad classes of elements are defined with regards to their contribution to a mineral's colour. Idiochromatic elements are essential to a mineral's composition; their contribution to a mineral's colour is diagnostic. Examples of such minerals are malachite (green) and azurite (blue). In contrast, allochromatic elements in minerals are present in trace amounts as impurities. An example of such a mineral would be the ruby and sapphire varieties of the mineral corundum. The colours of pseudochromatic minerals are the result of interference of light waves. Examples include labradorite and bornite.

In addition to simple body colour, minerals can have various other distinctive optical properties, such as play of colours, asterism, chatoyancy, iridescence, tarnish, and pleochroism. Several of these properties involve variability in colour. Play of colour, such as in opal, results in the sample reflecting different colours as it is turned, while pleochroism describes the change in colour as light passes through a mineral in a different orientation. Iridescence is a variety of the play of colours where light scatters off a coating on the surface of crystal, cleavage planes, or off layers having minor gradations in chemistry. In contrast, the play of colours in opal is caused by light refracting from ordered microscopic silica spheres within its physical structure. Chatoyancy ("cat's eye") is the wavy banding of colour that is observed as the sample is rotated; asterism, a variety of chatoyancy, gives the appearance of a star on the mineral grain. The latter property is particularly common in gem-quality corundum.

The streak of a mineral refers to the colour of a mineral in powdered form, which may or may not be identical to its body colour. The most common way of testing this property is done with a streak plate, which is made out of porcelain and coloured either white or black. The streak of a mineral is independent of trace elements or any weathering surface. A common example of this property is illustrated with hematite, which is coloured black, silver, or red in hand sample, but has a cherry-red to reddish-brown streak. Streak is more often distinctive for metallic minerals, in contrast to non-metallic minerals whose body colour is created by allochromatic elements. Streak testing is constrained by the hardness of the mineral, as those harder than 7 powder the *streak plate* instead.

Cleavage, Parting, Fracture, and Tenacity

Perfect basal cleavage as seen in biotite (black), and good cleavage seen in the matrix (pink orthoclase).

By definition, minerals have a characteristic atomic arrangement. Weakness in this crystalline structure causes planes of weakness, and the breakage of a mineral along such planes is termed cleavage. The quality of cleavage can be described based on how cleanly and easily the mineral breaks; common descriptors, in order of decreasing quality, are "perfect", "good", "distinct", and "poor". In particularly transparent mineral, or in thin-section, cleavage can be seen a series of parallel lines marking the planar surfaces when viewed at a side. Cleavage is not a universal property among minerals; for example, quartz, consisting of extensively interconnected silica tetrahedra, does not have a crystallographic weakness which would allow it to cleave. In contrast, micas, which have perfect basal cleavage, consist of sheets of silica tetrahedra which are very weakly held together.

As cleavage is a function of crystallography, there are a variety of cleavage types. Cleavage occurs typically in either one, two, three, four, or six directions. Basal cleavage in one direction is a distinctive property of the micas. Two-directional cleavage is described as prismatic, and occurs in minerals such as the amphiboles and pyroxenes. Minerals such as galena or halite have cubic (or isometric) cleavage in three directions, at 90°; when three directions of cleavage are present, but not at 90°, such as in calcite or rhodochrosite, it is termed rhombohedral cleavage. Octahedral cleavage (four directions) is present in fluorite and diamond, and sphalerite has six-directional dodecahedral cleavage.

Minerals with many cleavages might not break equally well in all of the directions; for example, calcite has good cleavage in three direction, but gypsum has perfect cleavage in one direction, and poor cleavage in two other directions. Angles between cleavage planes vary between minerals. For example, as the amphiboles are double-chain silicates and the pyroxenes are single-chain silicates, the angle between their cleavage planes is different. The pyroxenes cleave in two directions at approximately 90°, whereas the amphiboles distinctively cleave in two directions separated by approximately 120° and 60°. The cleavage angles can be measured with a contact goniometer, which is similar to a protractor.

Parting, sometimes called "false cleavage", is similar in appearance to cleavage but is instead produced by structural defects in the mineral as opposed to systematic weakness. Parting varies from crystal to crystal of a mineral, whereas all crystals of a given mineral will cleave if the atomic structure allows for that property. In general, parting is caused by some stress applied to a crystal. The sources of the stresses include deformation (e.g. an increase in pressure), exsolution, or twinning. Minerals that often display parting include the pyroxenes, hematite, magnetite, and corundum.

When a mineral is broken in a direction that does not correspond to a plane of cleavage, it is termed to have been fractured. There are several types of uneven fracture. The classic example is conchoidal fracture, like that of quartz; rounded surfaces are created, which are marked by smooth curved lines. This type of fracture occurs only in very homogeneous minerals. Other types of fracture are fibrous, splintery, and hackly. The latter describes a break along a rough, jagged surface; an example of this property is found in native copper.

Tenacity is related to both cleavage and fracture. Whereas fracture and cleavage describes the surfaces that are created when a mineral is broken, tenacity describes how resistant a mineral is to such breaking. Minerals can be described as brittle, ductile, malleable, sectile, flexible, or elastic.

Specific Gravity

Galena, PbS, is a mineral with a high specific gravity.

Specific gravity numerically describes the density of a mineral. The dimensions of density are mass divided by volume with units: kg/m^3 or g/cm^3. Specific gravity measures how much water a mineral sample displaces. Defined as the quotient of the mass of the sample and difference between the weight of the sample in air and its corresponding weight in water, specific gravity is a unitless ratio. Among most minerals, this property is not diagnostic. Rock forming minerals — typically silicates or occasionally carbonates — have a specific gravity of 2.5–3.5.

High specific gravity is a diagnostic property of a mineral. A variation in chemistry (and consequently, mineral class) correlates to a change in specific gravity. Among more common minerals, oxides and sulfides tend to have a higher specific gravity as they include elements with higher atomic mass. A generalization is that minerals with metallic or adamantine lustre tend to have higher specific gravities than those having a non-metallic to dull lustre. For example, hematite, Fe_2O_3, has a specific gravity of 5.26 while galena, PbS, has a specific gravity of 7.2–7.6, which is a result of their high iron and lead content, respectively. A very high specific gravity becomes very pronounced in native metals; kamacite, an iron-nickel alloy common in iron meteorites has a specific gravity of 7.9, and gold has an observed specific gravity between 15 and 19.3.

Other Properties

Carnotite (yellow) is a radioactive uranium-bearing mineral.

Other properties can be used to diagnose minerals. These are less general, and apply to specific minerals.

Dropping dilute acid (often 10% HCl) aids in distinguishing carbonates from other mineral classes. The acid reacts with the carbonate ($[CO_3]^{2-}$) group, which causes the affected area to effervesce, giving off carbon dioxide gas. This test can be further expanded to test the mineral in its original crystal form or powdered. An example of this test is done when distinguish calcite from dolomite, especially within rocks (limestone and dolostone respectively). Calcite immediately effervesces in acid, whereas acid must be applied to powdered dolomite (often to a scratched surface in a rock), for it to effervesce. Zeolite minerals will not effervesce in acid; instead, they become frosted after 5–10 minutes, and if left in acid for a day, they dissolve or become a silica gel.

When tested, magnetism is a very conspicuous property of minerals. Among common minerals, magnetite exhibits this property strongly, and it is also present, albeit not as strongly, in pyrrhotite and ilmenite.

Minerals can also be tested for taste or smell. Halite, NaCl, is table salt; its potassium-bearing counterpart, sylvite, has a pronounced bitter taste. Sulfides have a characteristic smell, especially as samples are fractured, reacting, or powdered.

Radioactivity is a rare property; minerals may be composed of radioactive elements. They could be a defining constituent, such as uranium in uraninite, autunite, and carnotite, or as trace impurities. In the latter case, the decay of a radioactive element damages the mineral crystal; the result, termed a *radioactive halo* or *pleochroic halo*, is observable by various techniques, such as thin-section petrography.

Mineral Classes

As the composition of the Earth's crust is dominated by silicon and oxygen, silicate elements are by far the most important class of minerals in terms of rock formation and diversity. However, non-silicate minerals are of great economic importance, especially as ores.

Non-silicate minerals are subdivided into several other classes by their dominant chemistry, which included native elements, sulfides, halides, oxides and hydroxides, carbonates and nitrates, borates, sulfates, phosphates, and organic compounds. The majority of non-silicate mineral species are extremely rare (constituting in total 8% of the Earth's crust), although some are relative common, such as calcite, pyrite, magnetite, and hematite. There are two major structural styles observed in non-silicates: close-packing and silicate-like linked tetrahedra. The close-packed structures, which is a way to densely pack atoms while minimizing interstitial space. Hexagonal close-packing involves stacking layers where every other layer is the same ("ababab"), whereas cubic close-packing involves stacking groups of three layers ("abcabcabc"). Analogues to linked silica tetrahedra include SO_4 (sulfate), PO_4 (phosphate), AsO_4 (arsenate), and VO_4 (vanadate). The non-silicates have great economic importance, as they concentrate elements more than the silicate minerals do.

The largest grouping of minerals by far are the silicates; most rocks are composed of greater than 95% silicate minerals, and over 90% of the Earth's crust is composed of these minerals. The two main constituents of silicates are silicon and oxygen, which are the two most abundant elements in the Earth's crust. Other common elements in silicate minerals correspond to other common elements in the Earth's crust, such aluminium, magnesium, iron, calcium, sodium, and potassium. Some important rock-forming silicates include the feldspars, quartz, olivines, pyroxenes, amphiboles, garnets, and micas.

Silicates

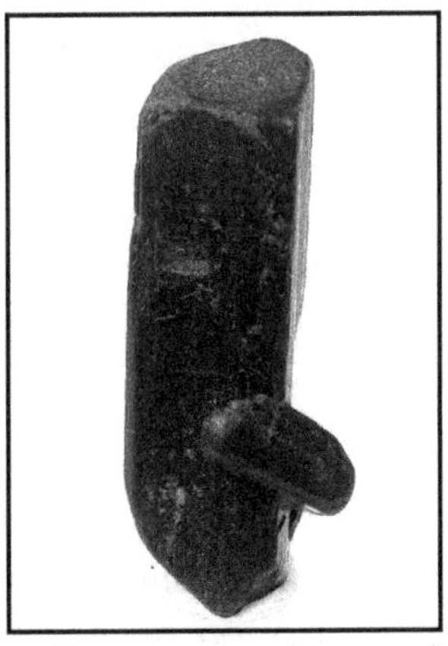

Aegirine, an iron-sodium clinopyroxene, is part of the inosilicate subclass.

The base of unit of a silicate mineral is the $[SiO_4]^{4-}$ tetrahedron. In the vast majority of cases, silicon is in four-fold or tetrahedral coordination with oxygen. In very high-pressure situations, silicon will be six-fold or octahedral coordination, such as in the perovskite structure or the quartz polymorph stishovite (SiO_2). In the latter case, the mineral no longer has a silicate structure, but that of rutile (TiO_2), and its associated group, which are simple oxides. These silica tetrahedra are then polymerized to some degree to create various structures, such as one-dimensional chains, two-dimensional sheets, and three-dimensional frameworks. The basic silicate mineral where no polymerization of the tetrahedra has occurred requires other elements to balance out the base 4- charge. In other silicate structures, different combinations of elements are required to balance out the resultant negative charge. It is common for the Si^{4+} to be substituted by Al^{3+} because of similarity in ionic radius and charge; in those case, the $[AlO_4]^{5-}$ tetrahedra form the same structures as do the unsubstituted tetrahedra, but their charge-balancing requirements are different.

The degree of polymerization can be described by both the structure formed and how many tetrahedral corners (or coordinating oxygens) are shared (for aluminium and silicon in tetrahedral sites). Orthosilicates (or nesosilicates) have no linking of polyhedra, thus tetrahedra share no corners. Disilicates (or sorosilicates) have two tetrahedra sharing one oxygen atom. Inosilicates are chain silicates; single-chain silicates have two shared corners, whereas double-chain silicates have two or three shared corners. In phyllosilicates, a sheet structure is formed which requires three shared oxygens; in the case of double-chain silicates, some tetrahedra must share two corners instead of three as otherwise a sheet structure would result. Framework silicates, or tectosilicates, have tetrahedra that share all four corners. The ring silicates, or cyclosilicates, only need tetrahedra to share two corners to form the cyclical structure.

The silicate subclasses are described below in order of decreasing polymerization.

Tectosilicates

Natrolite is a mineral series in the zeolite group; this sample has a very prominent acicular crystal habit.

Tectosilicates, also known as framework silicates, have the highest degree of polymerization. With all corners of a tetrahedra shared, the silicon:oxygen ratio becomes 1:2. Examples are quartz, the feldspars, feldspathoids, and the zeolites. Framework silicates tend to be particularly chemically stable as a result of strong covalent bonds.

Forming 12% of the Earth's crust, quartz (SiO_2) is the most abundant mineral species. It is characterized by its high chemical and physical resistivity. Quartz has several polymorphs, including tridymite and cristobalite at high temperatures, high-pressure coesite, and ultra-high pressure stishovite. The latter mineral can only be formed on Earth by meteorite impacts, and its structure has been composed so much that it had changed from a silicate structure to that of rutile

(TiO_2). The silica polymorph that is most stable at the Earth's surface is α-quartz. Its counterpart, β-quartz, is present only at high temperatures and pressures (changes to α-quartz below 573 °C at 1 bar). These two polymorphs differ by a "kinking" of bonds; this change in structure gives β-quartz greater symmetry than α-quartz, and they are thus also called high quartz (β) and low quartz (α).

Feldspars are the most abundant group in the Earth's crust, at about 50%. In the feldspars, Al^{3+} substitutes for Si^{4+}, which creates a charge imbalance that must be accounted for by the addition of cations. The base structure becomes either $[AlSi_3O_8]^-$ or $[Al_2Si_2O_8]^{2-}$ There are 22 mineral species of feldspars, subdivided into two major subgroups—alkali and plagioclase—and two less common groups—celsian and banalsite. The alkali feldspars are most commonly in a series between potassium-rich orthoclase and sodium-rich albite; in the case of plagioclase, the most common series ranges from albite to calcium-rich anorthite. Crystal twinning is common in feldspars, especially polysynthetic twins in plagioclase and Carlsbad twins in alkali feldspars. If the latter subgroup cools slowly from a melt, it forms exsolution lamellae because the two components—orthoclase and albite—are unstable in solid solution. Exsolution can be on a scale from microscopic to readily observable in hand-sample; perthitic texture forms when Na-rich feldspar exsolve in a K-rich host. The opposite texture (antiperthitic), where K-rich feldspar exsolves in a Na-rich host, is very rare.

Feldspathoids are structurally similar to feldspar, but differ in that they form in Si-deficient conditions which allows for further substitution by Al^{3+}. As a result, feldsapthoids cannot be associated with quartz. A common example of a feldsapthoid is nepheline ($(Na, K)AlSiO_4$); compared to alkali feldspar, nepheline has an Al_2O_3:SiO_2 ratio of 1:2, as opposed to 1:6 in the feldspar. Zeolites often have distinctive crystal habits, occurring in needles, plates, or blocky masses. They form in the presence of water at low temperatures and pressures, and have channels and voids in their structure. Zeolites have several industrial applications, especially in waste water treatment.

Phyllosilicates

Muscovite, a mineral species in the mica group, within the phyllosilicate subclass.

Phyllosilicates consist of sheets of polymerized tetrahedra. They are bound at three oxygen sites, which gives a characteristic silicon:oxygen ratio of 2:5. Important examples include the mica, chlorite, and the kaolinite-serpentine groups. The sheets are weakly bound by van der Waals forces or hydrogen bonds, which causes a crystallographic weakness, in turn leading to a prominent basal cleavage among the phyllosilicates. In addition to the tetrahedra, phyllosilicates have a sheet of octahedra (elements in six-fold coordination by oxygen) that balanced out the basic tetrahedra, which have a negative charge (e.g. $[Si_4O_{10}]^{4-}$) These tetrahedra (T) and octahedra (O) sheets are stacked in a variety of combinations to create phyllosilicate groups. Within an octahedral sheet, there are three octahedral sites in a unit structure; however, not all of the sites may be occupied. In that case, the mineral is termed dioctahedral, whereas in other case it is termed trioctahedral.

The kaolinite-serpentine group consists of T-O stacks (the 1:1 clay minerals); their hardness ranges from 2 to 4, as the sheets are held by hydrogen bonds. The 2:1 clay minerals (pyrophyllite-talc) consist of T-O-T stacks, but they are softer (hardness from 1 to 2), as they are instead held together by van der Waals forces. These two groups of minerals are subgrouped by octahedral occupation; specifically, kaolinite and pyrophyllite are dioctahedral whereas serpentine and talc trioctahedral.

Micas are also T-O-T-stacked phyllosilicates, but differ from the other T-O-T and T-O-stacked subclass members in that they incorporate aluminium into the tetrahedral sheets (clay minerals have Al^{3+} in octahedral sites). Common examples of micas are muscovite, and the biotite series. The chlorite group is related to mica group, but a brucite-like ($Mg(OH)_2$) layer between the T-O-T stacks.

Because of their chemical structure, phyllosilicates typically have flexible, elastic, transparent layers that are electrical insulators and can be split into very thin flakes. Micas can be used in electronics as insulators, in construction, as optical filler, or even cosmetics. Chrysotile, a species of serpentine, is the most common mineral species in industrial asbestos, as it is less dangerous in terms of health than the amphibole asbestos.

Inosilicates

Asbestiform tremolite, part of the amphibole group in the inosilicate subclass.

Inosilicates consist of tetrahedra repeatedly bonded in chains. These chains can be single, where a tetrahedron is bound to two others to form a continuous chain; alternatively, two chains can be merged to create double-chain silicates. Single-chain silicates have a silicon:oxygen ratio of 1:3 (e.g. $[Si_2O_6]^{4-}$), whereas the double-chain variety has a ratio of 4:11, e.g. $[Si_8O_{22}]^{12-}$. Inosilicates contain two important rock-forming mineral groups; single-chain silicates are most commonly pyroxenes, while double-chain silicates are often amphiboles. Higher-order chains exist (e.g. three-member, four-member, five-member chains, etc.) but they are rare.

The pyroxene group consists of 21 mineral species. Pyroxenes have a general structure formula of $XY(Si_2O_6)$, where X is an octahedral site, while Y can vary in coordination number from six to eight.

Most varieties of pyroxene consist of permutations of Ca^{2+}, Fe^{2+} and Mg^{2+} to balance the negative charge on the backbone. Pyroxenes are common in the Earth's crust (about 10%) and are a key constituent of mafic igneous rocks.

Amphiboles have great variability in chemistry, described variously as a "mineralogical garbage can" or a "mineralogical shark swimming a sea of elements". The backbone of the amphiboles is the $[Si_8O_{22}]^{12-}$; it is balanced by cations in three possible positions, although the third position is not always used, and one element can occupy both remaining ones. Finally, the amphiboles are usually hydrated, that is, they have a hydroxyl group ($[OH]^-$), although it can be replaced by a fluoride, a chloride, or an oxide ion. Because of the variable chemistry, there are over 80 species of amphibole, although variations, as in the pyroxenes, most commonly involve mixtures of Ca^{2+}, Fe^{2+} and Mg^{2+}. Several amphibole mineral species can have an asbestiform crystal habit. These asbestos minerals form long, thin, flexible, and strong fibres, which are electrical insulators, chemically inert and heat-resistant; as such, they have several applications, especially in construction materials. However, asbestos are known carcinogens, and cause various other illnesses, such as asbestosis; amphibole asbestos (anthophyllite, tremolite, actinolite, grunerite, and riebeckite) are considered more dangerous than chrysotile serpentine asbestos.

Cyclosilicates

An example of elbaite, a species of tourmaline, with distinctive colour banding.

Cyclosilicates, or ring silicates, have a ratio of silicon to oxygen of 1:3. Six-member rings are most common, with a base structure of $[Si_6O_{18}]^{12-}$; examples include the tourmaline group and beryl. Other ring structures exist, with 3, 4, 8, 9, 12 having been described. Cyclosilicates tend to be strong, with elongated, striated crystals.

Tourmalines have a very complex chemistry that can be described by a general formula $XY_3Z_6(BO_3)_3T_6O_{18}V_3W$. The T_6O_{18} is the basic ring structure, where T is usually Si^{4+}, but substitutable by Al_{3+} or B^{3+}. Tourmalines can be subgrouped by the occupancy of the X site, and from there further subdivided by the chemistry of the W site. The Y and Z sites can accommodate a variety of cations, especially various transition metals; this variability in structural transition metal content gives the tourmaline group greater variability in colour. Other cyclosilicates include beryl, $Al_2Be_3Si_6O_{18}$, whose varieties include the gemstones emerald (green) and aquamarine (bluish). Cordierite is structurally similar to beryl, and is a common metamorphic mineral.

Sorosilicates

Sorosilicates, also termed disilicates, have tetrahedron-tetrahedron bonding at one oxygen, which results in a 2:7 ratio of silicon to oxygen. The resultant common structural element is the $[Si_2O_7]^{6-}$ group. The most common disilicates by far are members of the epidote group. Epidotes are found in variety of geologic settings, ranging from mid-ocean ridge to granites to metapelites. Epidotes are built around the structure $[(SiO_4)(Si_2O_7)]^{10-}$ structure; for example, the mineral *species* epidote has calcium, aluminium, and ferric iron to charge balance: $Ca_2Al_2(Fe^{3+}, Al)(SiO_4)(Si_2O_7)O(OH)$. The presence of iron as Fe^{3+} and Fe^{2+} helps understand oxygen fugacity, which in turn is a significant factor in petrogenesis.

Other examples of sorosilicates include lawsonite, a metamorphic mineral forming in the blueschist facies (subduction zone setting with low temperature and high pressure), vesuvianite, which takes up a significant amount of calcium in its chemical structure.

Orthosilicates

Black andradite, an end-member of the orthosilicate garnet group.

Orthosilicates consist of isolated tetrahedra that are charge-balanced by other cations. Also termed nesosilicates, this type of silicate has a silicon:oxygen ratio of 1:4 (e.g. SiO_4). Typical orthosilicates tend to form blocky equant crystals, and are fairly hard. Several rock-forming minerals are part of this subclass, such as the aluminosilicates, the olivine group, and the garnet group.

The aluminosilicates—kyanite, andalusite, and sillimanite, all Al_2SiO_5—are structurally composed of one $[SiO_4]^{4-}$ tetrahedron, and one Al^{3+} in octahedral coordination. The remaining Al^{3+} can be in six-fold coordination (kyanite), five-fold (andalusite) or four-fold (sillimanite); which mineral forms in a given environment is depend on pressure and temperature conditions. In the olivine structure, the main olivine series of $(Mg, Fe)_2SiO_4$ consist of magnesium-rich forsterite and iron-rich fayalite. Both iron and magnesium are in octahedral by oxygen. Other mineral species having this structure exist, such as tephroite, Mn_2SiO_4. The garnet group has a general formula of $X_3Y_2(SiO_4)_3$, where X is a large eight-fold coordinated cation, and Y is a smaller six-fold coordinated cation. There are six ideal endmembers of garnet, split into two group. The pyralspite garnets have Al^{3+} in the Y position: pyrope $(Mg_3Al_2(SiO_4)_3)$, almandine $(Fe_3Al_2(SiO_4)_3)$, and spessartine $(Mn_3Al_2(SiO_4)_3)$. The ugrandite garnets have Ca^{2+} in the X position: uvarovite $(Ca_3Cr_2(SiO_4)_3)$, grossular $(Ca_3Al_2(SiO_4)_3)$ and andradite $(Ca_3Fe_2(SiO_4)_3)$. While there are two subgroups of garnet, solid solutions exist between all six end-members.

Other orthosilicates include zircon, staurolite, and topaz. Zircon ($ZrSiO_4$) is useful in geochronology as the Zr^{4+} can be substituted by U^{6+}; furthermore, because of its very resistant structure, it is difficult to reset it as a chronometer. Staurolite is a common metamorphic intermediate-grade index mineral. It has a particularly complicated crystal structure that was only fully described in 1986. Topaz ($Al_2SiO_4(F, OH)_2$, often found in granitic pegmatites associated with tourmaline, is a common gemstone mineral.

Non-Silicates

Native Elements

Native gold. Rare specimen of stout crystals growing off of a central stalk, size 3.7 x 1.1 x 0.4 cm, from Venezuela.

Native elements are those that are not chemically bonded to other elements. This mineral group includes native metals, semi-metals, and non-metals, and various alloys and solid solutions. The metals are held together by metallic bonding, which confers distinctive physical properties such as their shiny metallic lustre, ductility and malleability, and electrical conductivity. Native elements are subdivided into groups by their structure or chemical attributes.

The gold group, with a cubic close-packed structure, includes metals such as gold, silver, and copper. The platinum group is similar in structure to the gold group. The iron-nickel group is characterized by several iron-nickel alloy species. Two examples are kamacite and taenite, which are found in iron meteorites; these species differ by the amount of Ni in the alloy; kamacite has less than 5–7% nickel and is a variety of native iron, whereas the nickel content of taenite ranges from 7–37%. Arsenic group minerals consist of semi-metals, which have only some metallic; for example, they lack the malleability of metals. Native carbon occurs in two allotropes, graphite and diamond; the latter forms at very high pressure in the mantle, which gives it a much stronger structure than graphite.

Sulfides

The sulfide minerals are chemical compounds of one or more metals or semimetals with a sulfur; tellurium, arsenic, or selenium can substitute for the sulfur. Sulfides tend to be soft, brittle minerals with a high specific gravity. Many powdered sulfides, such as pyrite, have a sulfurous smell when powdered. Sulfides are susceptible to weathering, and many readily dissolve in water; these dissolved minerals can be later redeposited, which creates enriched secondary ore deposits. Sulfides are classified by the ratio of the metal or semimetal to the sulfur, such as M:S equal to 2:1, or

1:1. Many sulfide minerals are economically important as metal ores; examples include sphalerite (ZnS), an ore of zinc, galena (PbS), an ore of lead, cinnabar (HgS), an ore of mercury, and molybdenite (MoS_2, an ore of molybdenum. Pyrite (FeS_2), is the most commonly occurring sulfide, and can be found in most geological environments. It is not, however, an ore of iron, but can be instead oxidized to produce sulfuric acid. Related to the sulfides are the rare sulfosalts, in which a metallic element is bonded to sulfur and a semimetal such as antimony, arsenic, or bismuth. Like the sulfides, sulfosalts are typically soft, heavy, and brittle minerals.

Oxides

Oxide minerals are divided into three categories: simple oxides, hydroxides, and multiple oxides. Simple oxides are characterized by O^{2-} as the main anion and primarily ionic bonding. They can be further subdivided by the ratio of oxygen to the cations. The periclase group consists of minerals with a 1:1 ratio. Oxides with a 2:1 ratio include cuprite (Cu_2O) and water ice. Corundum group minerals have a 2:3 ratio, and includes minerals such as corundum (Al_2O_3), and hematite (Fe_2O_3). Rutile group minerals have a ratio of 1:2; the eponymous species, rutile (TiO_2) is the chief ore of titanium; other examples include cassiterite (SnO_2; ore of tin), and pyrolusite (MnO_2; ore of manganese). In hydroxides, the dominant anion is the hydroxyl ion, OH^-. Bauxites are the chief aluminium ore, and are a heterogeneous mixture of the hydroxide minerals diaspore, gibbsite, and bohmite; they form in areas with a very high rate of chemical weathering (mainly tropical conditions). Finally, multiple oxides are compounds of two metals with oxygen. A major group within this class are the spinels, with a general formula of $X^{2+}Y^{3+}{}_2O_4$. Examples of species include spinel ($MgAl_2O_4$), chromite ($FeCr_2O_4$), and magnetite (Fe_3O_4). The latter is readily distinguishable by its strong magnetism, which occurs as it has iron in two oxidation states ($Fe^{2+}Fe^{3+}{}_2O_4$), which makes it a multiple oxide instead of a single oxide.

Halides

Pink cubic halite (NaCl; halide class) crystals on a nahcolite matrix ($NaHCO_3$; a carbonate, and mineral form of sodium bicarbonate, used as baking soda).

The halide minerals are compounds where a halogen (fluorine, chlorine, iodine, and bromine) is the main anion. These minerals tend to be soft, weak, brittle, and water-soluble. Common examples of halides include halite (NaCl, table salt), sylvite (KCl), fluorite (CaF_2). Halite and sylvite commonly form as evaporites, and can be dominant minerals in chemical sedimentary rocks. Cryolite, Na_3AlF_6, is a key mineral in the extraction of aluminium from bauxites; however, as the only significant occurrence at Ivittuut, Greenland, in a granitic pegmatite, was depleted, synthetic cryolite can be made from fluorite.

Carbonates

The carbonate minerals are those were the main anionic group is carbonate, $[CO_3]^{2-}$. Carbonates tend to be brittle, many have rhombohedral cleavage, and all react with acid. Due to the last characteristic, field geologists often carry dilute hydrochloric acid to distinguish carbonates from non-carbonates. The reaction of acid with carbonates, most commonly found as the polymorph calcite and aragonite ($CaCO_3$), relates to the dissolution and precipitation of the mineral, which is a key in the formation of limestone caves, features within them such as stalactite and stalagmites, and karst landforms. Carbonates are most often formed as biogenic or chemical sediments in marine environments. The carbonate group is structurally a triangle, where a central C^{4+} cation is surrounded by three O^{2-} anions; different groups of minerals form from different arrangements of these triangles. The most common carbonate mineral is calcite, and is the primary constituent of sedimentary limestone and metamorphic marble. Calcite, $CaCO_3$, can have a high magnesium impurity; under high-Mg conditions, its polymorph aragonite will form instead; the marine geochemistry in this regard can be described as an aragonite or calcite sea, depending on which mineral preferentially forms. Dolomite is a double carbonate, with the formula $CaMg(CO_3)_2$. Secondary dolomitization of limestone is common, where calcite or aragonite are converted to dolomite; this reaction increases pore space (the unit cell volume of dolomite is 88% that of calcite), which can create a reservoir for oil and gas. These two minerals species are members of eponymous mineral groups: the calcite group includes carbonates with the general formula XCO_3, and the dolomite group constitutes minerals with general formula $XY(CO_3)_2$.

Sulfates

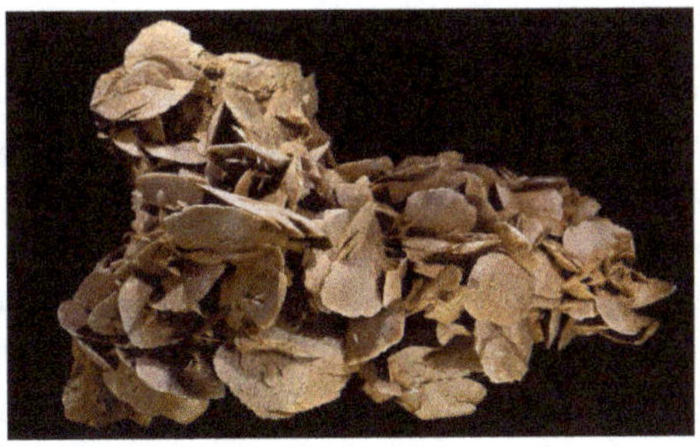

Gypsum desert rose

The sulfate minerals all contain the sulfate anion, $[SO_4]^{2-}$. They tend to be transparent to translucent, soft, and many are fragile. Sulfate minerals commonly form as evaporites, where they precipitate out of evaporating saline waters; alternative, sulfates can also be found in hydrothermal vein systems associated with sulfides, or as oxidation products of sulfides. Sulfates can be subdivided into anhydrous and hydrous minerals. The most common hydrous sulfate by far is gypsum, $CaSO_4{\cdot}2H_2O$. It forms as an evaporite, and is associated with other evaporites such as calcite and halite; if it incorporates sand grains as it crystallizes, gypsum can form desert roses. Gypsum has very low thermal conductivity and maintains a low temperature when heated as it loses that heat by dehydrating; as such, gypsum is used as an insulator in materials such as plaster and drywall. The anhydrous equivalent of gypsum is anhydrite; it can form directly from seawater in highly arid conditions. The barite group has the general formula XSO_4, where the X is a large 12-coordinated cation. Examples include barite ($BaSO_4$), celestine ($SrSO_4$), and anglesite ($PbSO_4$); anhydrite is not part of the barite group, as the smaller Ca^{2+} is only in eight-fold coordination.

Phosphates

The phosphate minerals are characterized by the tetrahedral $[PO_4]^{3-}$ unit, although the structure can be generalized, and phosphorus is replaced by antimony, arsenic, or vanadium. The most common phosphate is the apatite group; common species within this group are fluorapatite ($Ca_5(PO_4)_3F$), chlorapatite ($Ca_5(PO_4)_3Cl$) and hydroxylapatite ($Ca_5(PO_4)_3(OH)$). Minerals in this group are the main crystalline constituents of teeth and bones in vertebrates. The relatively abundant monazite group has a general structure of ATO_4, where T is phosphorus or arsenic, and A is often a rare-earth element (REE). Monazite is important in two ways: first, as a REE "sink", it can sufficiently concentrate these elements to become an ore; secondly, monazite group elements can incorporate relatively large amounts of uranium and thorium, which can be used to date the rock based on the decay of the U and Th to lead.

Organic Minerals

The Strunz classification includes a class for organic minerals. These rare compounds contain organic carbon, but can be formed by a geologic process. For example, whewellite, $CaC_2O_4{\cdot}H_2O$ is an oxalate that can be deposited in hydrothermal ore veins. While hydrated calcium oxalate can be found in coal seams and other sedimentary deposits involving organic matter, the hydrothermal occurrence is not considered to be related to biological activity.

Astrobiology

It has been suggested that biominerals could be important indicators of extraterrestrial life and thus could play an important role in the search for past or present life on the planet Mars. Furthermore, organic components (biosignatures) that are often associated with biominerals are believed to play crucial roles in both pre-biotic and biotic reactions.

On January 24, 2014, NASA reported that current studies by the *Curiosity* and *Opportunity* rovers on Mars will now be searching for evidence of ancient life, including a biosphere based on autotrophic, chemotrophic and/or chemolithoautotrophic microorganisms, as well as ancient water, including fluvio-lacustrine environments (plains related to ancient rivers or lakes) that may have been habitable. The search for evidence of habitability, taphonomy (related to fossils), and organic carbon on the planet Mars is now a primary NASA objective.

Silicate Minerals

Silicate minerals are rock-forming minerals made up of silicate groups. They are the largest and most important class of rock-forming minerals and make up approximately 90 percent of the Earth's crust. They are classified based on the structure of their silicate groups, which contain different ratios of silicon and oxygen.

CNesosilicates or Orthosilicates

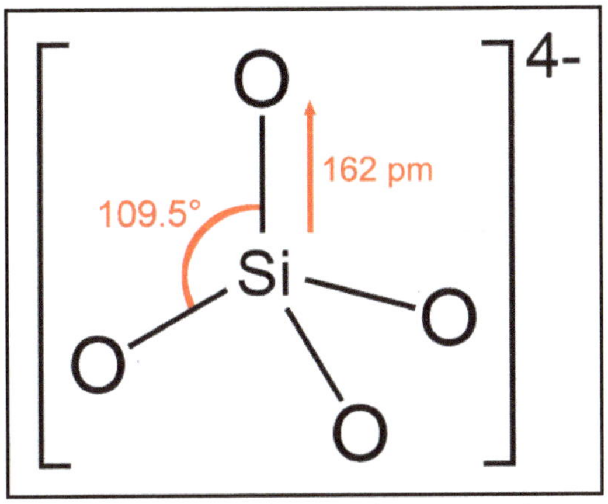

Basic (ortho-)silicate anion structure.

Nesosilicates or orthosilicates, have the orthosilicate ion, which constitute isolated (insular) $[SiO_4]^{4-}$ tetrahedra that are connected only by interstitial cations. Nickel–Strunz classification: 09.A

- Phenakite group:
    - Phenakite – Be_2SiO_4.
    - Willemite – Zn_2SiO_4.
- Olivine group:
    - Forsterite – Mg_2SiO_4.
    - Fayalite – Fe_2SiO_4.
    - Tephroite – Mn_2SiO_4.
- Garnet group:
    - Pyrope – $Mg_3Al_2(SiO_4)_3$.
    - Almandine – $Fe_3Al_2(SiO_4)_3$.
    - Spessartine – $Mn_3Al_2(SiO_4)_3$.
    - Grossular – $Ca_3Al_2(SiO_4)_3$.
    - Andradite – $Ca_3Fe_2(SiO_4)_3$.
    - Uvarovite – $Ca_3Cr_2(SiO_4)_3$
    - Hydrogrossular – $Ca_3Al_2Si_2O_8(SiO_4)_{3-m}(OH)_{4m}$.
- Zircon group:
    - Zircon – $ZrSiO_4$.
    - Thorite – $(Th,U)SiO_4$.

Kyanite crystals (unknown scale).

- Al_2SiO_5 group:
    - Andalusite – Al_2SiO_5.
    - Kyanite – Al_2SiO_5.
    - Sillimanite – Al_2SiO_5.
    - Dumortierite – $Al_{6.5-7}BO_3(SiO_4)_3(O,OH)_3$.
    - Topaz – $Al_2SiO_4(F,OH)_2$.
    - Staurolite – $Fe_2Al_9(SiO_4)_4(O,OH)_2$.
- Humite group – $(Mg,Fe)_7(SiO_4)_3(F,OH)_2$:
    - Norbergite – $Mg_3(SiO_4)(F,OH)_2$.
    - Chondrodite – $Mg_5(SiO_4)_2(F,OH)_2$.
    - Humite – $Mg_7(SiO_4)_3(F,OH)_2$.
    - Clinohumite – $Mg_9(SiO_4)_4(F,OH)_2$.
- Datolite – $CaBSiO_4(OH)$.
- Titanite – $CaTiSiO_5$.
- Chloritoid – $(Fe,Mg,Mn)_2Al_4Si_2O_{10}(OH)_4$.
- Mullite (aka Porcelainite) – $Al_6Si_2O_{13}$.

Sorosilicates

Sorosilicate exhibit at Museum of Geology in South Dakota.

Sorosilicates have isolated double tetrahedra groups with $(Si_2O_7)^{6-}$ or a ratio of 2:7. Nickel–Strunz classification: 09.B

- Hemimorphite (calamine) – $Zn_4(Si_2O_7)(OH)_2 \cdot H_2O$.
- Lawsonite – $CaAl_2(Si_2O_7)(OH)_2 \cdot H_2O$.
- Ilvaite – $CaFe^{II}{}_2Fe^{III}O(Si_2O_7)(OH)$.
- Epidote group (has both $(SiO_4)^{4-}$ and $(Si_2O_7)^{6-}$ groups):
    - Epidote – $Ca_2(Al,Fe)_3O(SiO_4)(Si_2O_7)(OH)$.
    - Zoisite – $Ca_2Al_3O(SiO_4)(Si_2O_7)(OH)$.
    - Clinozoisite – $Ca_2Al_3O(SiO_4)(Si_2O_7)(OH)$.
    - Tanzanite – $Ca_2Al_3O(SiO_4)(Si_2O_7)(OH)$.
    - Allanite – $Ca(Ce,La,Y,Ca)Al_2(Fe^{II},Fe^{III})O(SiO_4)(Si_2O_7)(OH)$.
    - Dollaseite-(Ce) – $CaCeMg_2AlSi_3O_{11}F(OH)$.
- Vesuvianite (idocrase) – $Ca_{10}(Mg,Fe)_2Al_4(SiO_4)_5(Si_2O_7)_2(OH)_4$.

Cyclosilicates

Cyclosilicate specimens at the Museum of Geology, South Dakota.

Cyclosilicates, or ring silicates, have linked tetrahedra with $(T_xO_{3x})^{2x-}$ or a ratio of 1:3. These exist as 3-member $(T_3O_9)^{6-}$ and 6-member $(T_6O_{18})^{12-}$ rings, where T stands for a tetrahedrally coordinated cation. Nickel–Strunz classification: 09.C.

- 3-member ring:
    - Benitoite – $BaTi(Si_3O_9)$.
- 6-member ring:
    - Axinite – $(Ca,Fe,Mn)_3Al_2(BO_3)(Si_4O_{12})(OH)$.
    - Beryl/Emerald – $Be_3Al_2(Si_6O_{18})$.
    - Sugilite – $KNa_2(Fe,Mn,Al)_2Li_3Si_{12}O_{30}$.
    - Cordierite – $(Mg,Fe)_2Al_3(Si_5AlO_{18})$.
    - Tourmaline – $(Na,Ca)(Al,Li,Mg)_{3-}(Al,Fe,Mn)_6(Si_6O_{18}(BO_3)_3(OH)_4$.

Note that the ring in axinite contains two B and four Si tetrahedra and is highly distorted compared to the other 6-member ring cyclosilicates:

- 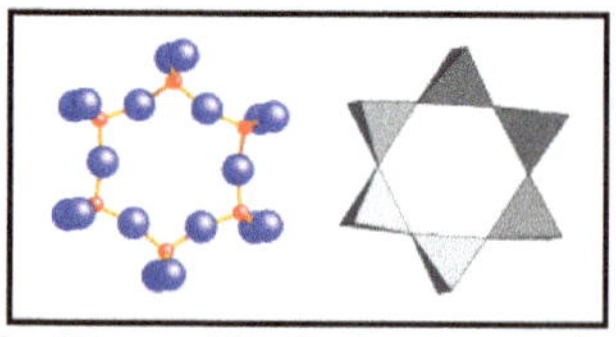

Cyclosilicate, $[Si_6O_{18}]$ – 6-membered single rings, beryl (red: Si, blue: O).

- 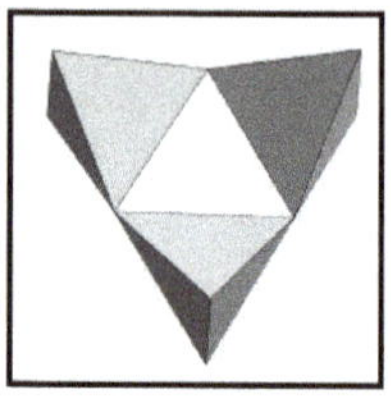

Cyclosilicate, $[Si_3O_9]$ – 3-membered single ring, benitoite.

- 

Cyclosilicate, $[Si_4O_{12}]$ – 4-membered single ring, papagoite.

- 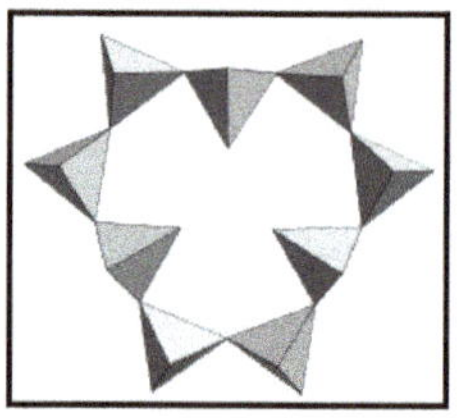

Cyclosilicate, $[Si_9O_{27}]$ – 9-membered ring, eudialyte.

- 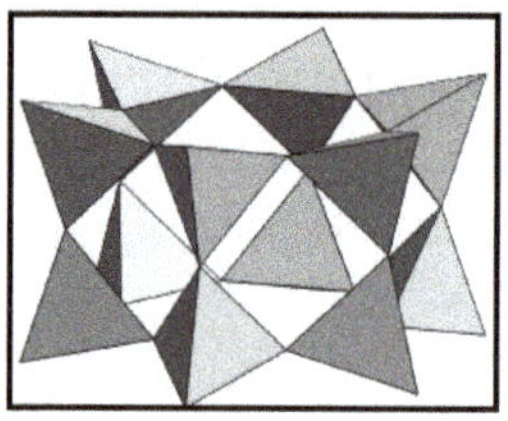

Cyclosilicate, $[Si_6O_{18}]$ – 6-membered double ring, milarite.

Inosilicates

Inosilicates, or chain silicates, have interlocking chains of silicate tetrahedra with either SiO_3, 1:3 ratio, for single chains or Si_4O_{11}, 4:11 ratio, for double chains. Nickel–Strunz classification: 09.D.

Single Chain Inosilicates

- Pyroxene group.

- Enstatite – orthoferrosilite series:
    - Enstatite – $MgSiO_3$.
    - Ferrosilite – $FeSiO_3$.
- Pigeonite – $Ca_{0.25}(Mg,Fe)_{1.75}Si_2O_6$.
- Diopside – hedenbergite series:
    - Diopside – $CaMgSi_2O_6$.
    - Hedenbergite – $CaFeSi_2O_6$.
    - Augite – $(Ca,Na)(Mg,Fe,Al)(Si,Al)_2O_6$.
- Sodium pyroxene series:
    - Jadeite – $NaAlSi_2O_6$.
    - Aegirine (Acmite) – $NaFe^{III}Si_2O_6$.
- Spodumene – $LiAlSi_2O_6$.

- Pyroxenoid group:
    - Wollastonite – $CaSiO_3$.
    - Rhodonite – $MnSiO_3$.
    - Pectolite – $NaCa_2(Si_3O_8)(OH)$.

Double Chain Inosilicates

- Amphibole group:
    - Anthophyllite – $(Mg,Fe)_7Si_8O_{22}(OH)_2$.
    - Cumingtonite series:
        - Cummingtonite – $Fe_2Mg_5Si_8O_{22}(OH)_2$.
        - Grunerite – $Fe_7Si_8O_{22}(OH)_2$.
    - Tremolite series:
        - Tremolite – $Ca_2Mg_5Si_8O_{22}(OH)_2$.
        - Actinolite – $Ca_2(Mg,Fe)_5Si_8O_{22}(OH)_2$.
    - Hornblende – $(Ca,Na)_{2-3}(Mg,Fe,Al)_5Si_6(Al,Si)_2O_{22}(OH)_2$.
    - Sodium amphibole group:
        - Glaucophane – $Na_2Mg_3Al_2Si_8O_{22}(OH)_2$.
        - Riebeckite (asbestos) – $Na_2Fe^{II}{}_3Fe^{III}{}_2Si_8O_{22}(OH)_2$.
        - Arfvedsonite – $Na_3(Fe,Mg)_4FeSi_8O_{22}(OH)_2$.

- 
  Inosilicate, pyroxene family, with 2-periodic single chain (Si_2O_6), diopside.
- 
  Inosilicate, clinoamphibole, with 2-periodic double chains (Si_4O_{11}), tremolite.
- 
  Inosilicate, unbranched 3-periodic single chain of wollastonite.
- 
  Inosilicate with 5-periodic single chain, rhodonite.

Phyllosilicates

Phyllosilicates, or sheet silicates, form parallel sheets of silicate tetrahedra with Si_2O_5 or a 2:5 ratio. Nickel–Strunz classification: 09.E. All phyllosilicate minerals are hydrated, with either water or hydroxyl groups attached.

Kaolin

- Serpentine subgroup:
    - Antigorite – $Mg_3Si_2O_5(OH)_4$.
    - Chrysotile – $Mg_3Si_2O_5(OH)_4$.
    - Lizardite – $Mg_3Si_2O_5(OH)_4$.
- Clay minerals group:
    - Halloysite – $Al_2Si_2O_5(OH)_4$.
    - Kaolinite – $Al_2Si_2O_5(OH)_4$.
    - Illite – $(K,H_3O)(Al,Mg,Fe)_2(Si,Al)_4O_{10}[(OH)_2,(H_2O)]$.
    - Montmorillonite – $(Na,Ca)_{0.33}(Al,Mg)_2Si_4O_{10}(OH)_2{\cdot}nH_2O$.
    - Vermiculite – $(MgFe,Al)_3(Al,Si)_4O_{10}(OH)_2{\cdot}4H_2O$.
    - Talc – $Mg_3Si_4O_{10}(OH)_2$.
    - Sepiolite – $Mg_4Si_6O_{15}(OH)_2{\cdot}6H_2O$.
    - Palygorskite (or attapulgite) – $(Mg,Al)_2Si_4O_{10}(OH){\cdot}4(H_2O)$.
    - Pyrophyllite – $Al_2Si_4O_{10}(OH)_2$.
- Mica group:
    - Biotite – $K(Mg,Fe)_3(AlSi_3)O_{10}(OH)_2$.
    - Muscovite – $KAl_2(AlSi_3)O_{10}(OH)_2$.
    - Phlogopite – $KMg_3(AlSi_3)O_{10}(OH)_2$.
    - Lepidolite – $K(Li,Al)_{2-3}(AlSi_3)O_{10}(OH)_2$.
    - Margarite – $CaAl_2(Al_2Si_2)O_{10}(OH)_2$.
    - Glauconite – $(K,Na)(Al,Mg,Fe)_2(Si,Al)_4O_{10}(OH)_2$.
- Chlorite group:
    - Chlorite – $(Mg,Fe)_3(Si,Al)_4O_{10}(OH)_2{\cdot}(Mg,Fe)_3(OH)_6$.

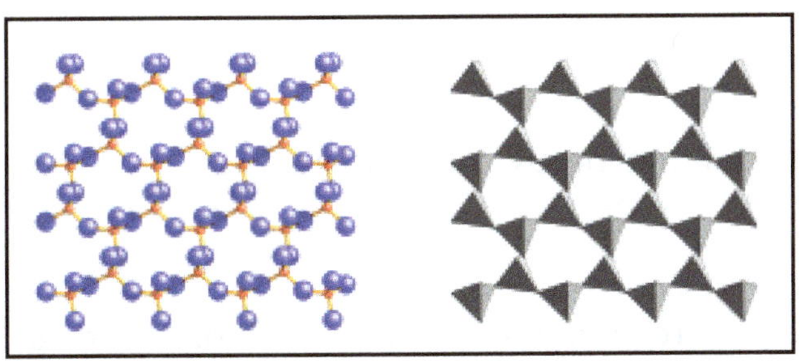

- Phyllosilicate, mica group, muscovite (red: Si, blue: O).

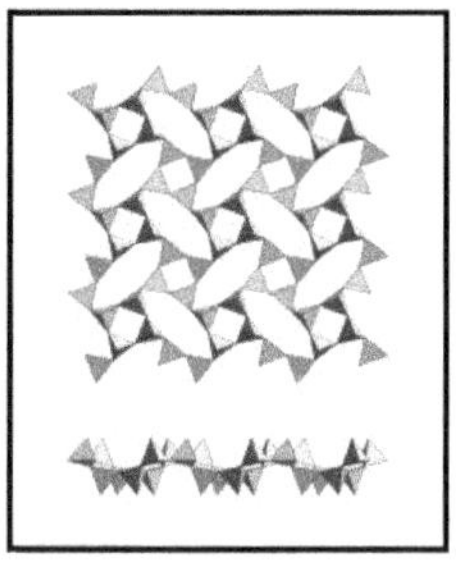

- Phyllosilicate, single net of tetrahedra with 4-membered rings, apophyllite-(KF)-apophyllite-(KOH) series.

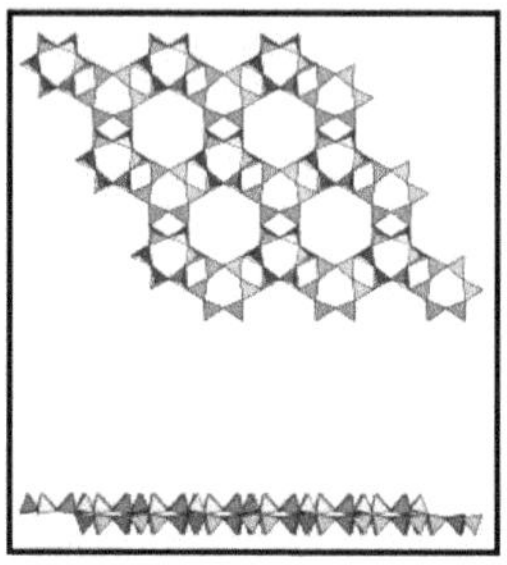

- Phyllosilicate, single tetrahedral nets of 6-membered rings, pyrosmalite-(Fe)-pyrosmalite-(Mn) series.

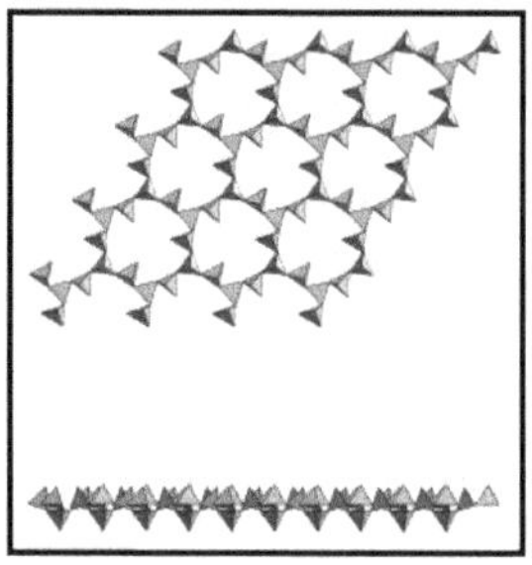

- Phyllosilicate, single tetrahedral nets of 6-membered rings, zeophyllite.

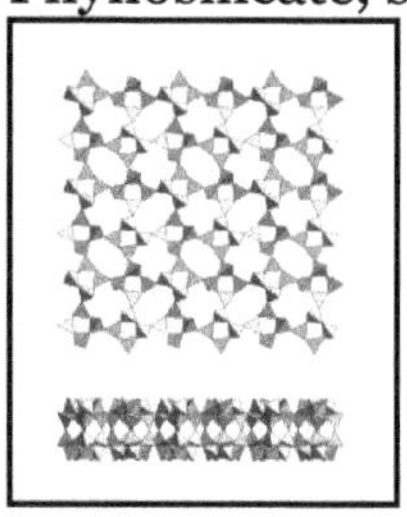

- Phyllosilicate, double nets with 4- and 6-membered rings, carletonite.

Tectosilicates

Tectosilicates, or "framework silicates," have a three-dimensional framework of silicate tetrahedra with SiO_2 or a 1:2 ratio. This group comprises nearly 75% of the crust of the Earth. Tectosilicates, with the exception of the quartz group, are aluminosilicates. Nickel–Strunz classification: 09.F and 09.G, 04.DA (Quartz/ silica family).

Lunar Ferroan Anorthosite no. 60025 (Plagioclase Feldspar). Collected by Apollo 16 from the Lunar Highlands near Descartes Crater.

- Quartz group:
  - Quartz – SiO_2.
  - Tridymite – SiO_2.
  - Cristobalite – SiO_2.
  - Coesite – SiO_2.
  - Stishovite – SiO_2.
- Feldspar family:
  - Alkali feldspars (potassium feldspars):
    - Microcline – $KAlSi_3O_8$.
    - Orthoclase – $KAlSi_3O_8$.
    - Anorthoclase – $(Na,K)AlSi_3O_8$.
    - Sanidine – $KAlSi_3O_8$.
    - Albite – $NaAlSi_3O_8$.
  - Plagioclase feldspars:
    - Albite – $NaAlSi_3O_8$.
    - Oligoclase – $(Na,Ca)(Si,Al)_4O_8$ (Na:Ca 4:1).
    - Andesine – $(Na,Ca)(Si,Al)_4O_8$ (Na:Ca 3:2).
    - Labradorite – $(Ca,Na)(Si,Al)_4O_8$ (Na:Ca 2:3).
    - Bytownite – $(Ca,Na)(Si,Al)_4O_8$ (Na:Ca 1:4).
    - Anorthite – $CaAl_2Si_2O_8$.
- Feldspathoid family:
    - Nosean – $Na_8Al_6Si_6O_{24}(SO_4)$.

        - Cancrinite – $Na_6Ca_2(CO_3,Al_6Si_6O_{24}).2H_2O$.
        - Leucite – $KAlSi_2O_6$.
        - Nepheline – $(Na,K)AlSiO_4$.
        - Sodalite – $Na_8(AlSiO_4)_6Cl_2$.
            - Hauyne – $(Na,Ca)_{4-8}Al_6Si_6(O,S)24(SO_4,Cl)_{1-2}$.
        - Lazurite – $(Na,Ca)_8(AlSiO_4)_6(SO_4,S,Cl)_2$.
    - Petalite – $LiAlSi_4O_{10}$.
    - Scapolite group:
        - Marialite – $Na_4(AlSi_3O_8)_3(Cl_2,CO_3,SO_4)$.
        - Meionite – $Ca_4(Al_2Si_2O_8)_3(Cl_2CO_3,SO_4)$.
    - Analcime – $NaAlSi_2O_6{\cdot}H_2O$.
    - Zeolite family:
        - Natrolite – $Na_2Al_2Si_3O_{10}{\cdot}2H_2O$.
        - Erionite – $(Na_2,K_2,Ca)_2Al_4Si_{14}O_{36}{\cdot}15H_2O$.
        - Chabazite – $CaAl_2Si_4O_{12}{\cdot}6H_2O$.
        - Heulandite – $CaAl_2Si_7O_{18}{\cdot}6H_2O$.
        - Stilbite – $NaCa_2Al_5Si_{13}O_{36}{\cdot}17H_2O$.
        - Scolecite – $CaAl_2Si_3O_{10}{\cdot}3H_2O$.
        - Mordenite – $(Ca,Na_2,K_2)Al_2Si_{10}O_{24}{\cdot}7H_2O$.

Classification of Silicate Minerals

This list gives an overview of the classification of minerals (silicates) and includes mostly IMA recognized minerals and its groupings. This list complements the alphabetical list on List of minerals (complete) and List of minerals. Rocks, ores, mineral mixtures, non-IMA approved minerals and non-named minerals are mostly excluded. Mostly major groups only, or groupings used by *New Dana Classification* and *Mindat*.

Classification of Minerals

Introduction

The grouping of the New Dana Classification and of the mindat.org is similar only, and so this classification is an overview only. Consistency is missing too on the group name endings (group, subgroup, series) between New Dana Classification and mindat.org. Category, class and supergroup name endings are used as layout tools in the list as well.

- Abbreviations:
    - "*" - mineral not IMA Approved.
    - "Q" - doubtful/ questionable.
    - Rn - renaming.
    - Rd - redefinition.
    - "REE" - rare earth element (Sc, Y, La, Ce, Pr, Nd, Pm, Sm, Eu, Gd, Tb, Dy, Ho, Er, Tm, Yb, Lu).
    - "PGE" - platinum group element (Ru, Rh, Pd, Os, Ir, Pt).
    - "s.p." - special procedure.

Category '9': Silicate Minerals

- Unclassified silicates.

Subclass '9.A': Nesosilicates

- Zircon group: $MSiO_4$ (a group of simple tetragonal silicates where M = tetravalent Zr, Th, Hf):
    - Hafnon $HfSiO_4$, Stetindite $Ce[SiO_4]$, Thorite $(Th,U)SiO_4$, Zircon $ZrSiO_4$, Coffinite $U(SiO_4)_{1-x}(OH)_{4x}$, Thorogummite $Th(SiO_4)_{1-x}(OH)_{4x}$, IMA2008-035 $CeSiO_4$.
- Olivine group:
    - Calcio-Olivine Ca_2SiO_4, Fayalite $(Fe^{2+})_2[SiO_4]$, Forsterite $Mg_2[SiO_4]$, Laihunite $Fe^{2+}(Fe^{3+})_2(SiO_4)_2$, Liebenbergite $(Ni,Mg)_2[SiO_4]$, Olivine $(Mg,Fe^{2+})_2[SiO_4]$, Tephroite $(Mn^{2+})_2[SiO_4]$.
- Phenakite group:
    - Phenakite Be_2SiO_4, Willemite Zn_2SiO_4, Eucryptite $LiAlSiO_4$.
- $Al_2(SiO_4)O$:
    - Sillimanite subgroup:
        - Sillimanite $AlAlOSiO_4$, Mullite $Al_{(4+2x)}Si_{(2-2x)}O_{(10-x)}$ (x =0.17 to 0.59), Boromullite $Al_{4.5}SiB_{0.5}O_{9.5}$.
    - Andalusite subgroup:
        - Andalusite $AlAlOSiO_4$, Kanonaite $(Mn^{3+},Al)AlSiO_5$, Yoderite $Mg_2(Al,Fe^{3+})_6Si_4O_{18}(OH)_2$.

  - Kyanite $AlAlOSiO_4$.
- Titanite group:
  - Titanite $CaTiSiO_5$, Malayaite $CaSnOSiO_4$, Vanadomalayaite $CaVOSiO_4$.
- Cerite Group:
  - Cerite-(Ce) $(Ce^{3+})_9Fe^{3+}(SiO_4)_6[(SiO_3)(OH)](OH)_3$, Cerite-(La) $(La,Ce,Ca)_9(Mg,Fe^{3+})(SiO_4)_6[SiO_3(OH)](OH)_3$, Aluminocerite-(Ce) $(Ce,REE,Ca)_9(Al,Fe^{3+})(SiO_4)_3[SiO_3(OH)]_4(OH)_3$.
- Silicate apatites:
  - Ellestadite* $Ca_5(SiO_4,PO_4,SO_4)_3(F,OH,Cl)$, Britholite-(Ce) $(Ce,Ca,Th,La,Nd)_5(SiO_4,PO_4)_3(OH,F)$, Britholite-(Y) $(Y,Ca)_5(SiO_4,PO_4)_3(OH,F)$, Ellestadite-(F) $Ca_5(SiO_4,PO_4,SO_4)_3(F,OH,Cl)$, Ellestadite-(OH) $Ca_5(SiO_4,SO_4)_3(OH,Cl,F)$, Ellestadite-(Cl) $Ca_5(SiO_4,PO_4,SO_4)_3(Cl,OH,F)$, Mattheddleite $Pb_2o(SiO_4)_7(SO_4)_4Cl_4$, Karnasurtite-(Ce) $(Ce,La,Th)(Ti,Nb)(Al,Fe^{3+})(Si,P)2O_7(OH)_4\cdot3(H_2O)$ (?), Fluorbritholite-(Ce) $(Ca,Ce,La,Na)_5(SiO_4,PO_4)_3(OH,F)$, Fluorcalciobritholite $(Ca,REE)_5[(Si,P)O_4]_3F$.
- Uranophane group:
  - Kasolite $Pb(UO_2)SiO_4\cdot H_2O$, Uranophane $Ca(UO_2)_2SiO_3(OH)_2\cdot5H_2O$, Sklodowskite $(H_3O)_2Mg(UO_2)_2(SiO_4)_2\cdot4H_2O$, Cuprosklodowskite $Cu[(UO_2)(SiO_2OH)]_2\cdot6H_2O$, Boltwoodite $HK(UO_2)(SiO_4)\cdot1.5H_2O$, Natroboltwoodite $(H_3O)(Na,K)(UO_2)SiO_4\cdot H_2O$, Oursinite $(Co,Mg)(H_3O)_2[(UO_2)SiO_4]_2\cdot3H_2O$, Swamboite $U^{7+}H_6(UO_2)_6(SiO_4)_6\cdot30H_2O$, Uranophane-beta $Ca[(UO_2)SiO_3(OH)]_2\cdot H_2O$.
- Datolite group:
  - Datolite series:
    - Datolite $CaBSiO_4(OH)$, Hingganite-(Ce) $(Ce,Ca)_2([\ ],Fe)Be_2Si_2O_8[(OH),O]_2$, Hingganite-(Y) $Y_2([\ ])Be_2Si_2O_8(OH)_2$, Hingganite-(Yb) $(Yb,Y)_2([\ ])Be_2Si_2O_8(OH)_2$, Calcybeborosilite-(Y) $(REE,Ca)_2[\ ](B,Be)_2(SiO_4)_2(OH,O)_2$.
- Homilite series:
    - Bakerite $Ca_4B_4(BO_4)(SiO_4)_3(OH)_3\cdot H_2O$, Gadolinite-(Ce) $(Ce,La,Nd,Y)_2Fe^{2+}Be_2Si_2O_{10}$, Gadolinite-(Y) $Y2Fe^{2+}Be_2Si_2O_{10}$, Calciogadolinite? $CaREE(Fe^{3+})Be_2Si_2O_{10}$, Homilite $Ca_2(Fe^{2+},Mg)B_2Si_2O_{10}$, Minasgeraisite-(Y) $CaY_2Be_2Si_2O_{10}$.
- Hellandite group:
  - Hellandite-(Y) $(Ca,REE)_4(Y,Ce)_2(Al,[\ ])_2[Si_4B_4O_{22}](OH)_2$.
  - Tadzhikite-(Y) $Ca_4(Y,Ce)_2(Ti,Al,Fe^{3+},[\ ])_2[Si_4B_4O_{22}](OH)_2$.
  - Tadzhikite-(Ce) $Ca_4(Ce,Y)_2(Ti,Al,Fe^{3+},[\ ])_2[Si_4B_4O_{22}](OH)_2$.
  - Hellandite-(Ce) $(Ca_3REE)_4Ce_2Al[\ ]_2[Si_4B_4O_{22}](OH)_2$.
  - Mottanaite-(Ce) Ca4(Ce,Ca)2AlBe2$[Si_4B_4O_{22}]O_2$.

- Ciprianiite $Ca_4[(Th,U)(REE)]_2(Al,[\ ])_2[Si_4B_4O_{22}](OH,F)_2$.
- Piergorite-(Ce) $Ca_8Ce_2(Al_{0.5}(Fe^{3+})_{0.5})([\ ],Li,Be)_2Si_6B_8O_{36}(OH,F)_2$.

- Vicanite group:
  - Vicanite-(Ce) $(Ca,Ce,La,Th)_{15}As^{5+}(As^{3+}{}_{0.5},Na_{0.5})Fe^{3+}Si_6B_4O_4oF_7$.
  - Hundholmenite-(Y) $(Y,REE,Ca,Na)_{15}(Al,Fe^{3+})Cax(As^{3+})_{1-x}(Si,As^{5+})Si_6B_3(O,F)_{48}$.
  - Proshchenkoite-(Y) $(Y,REE,Ca,Na,Mn)_{15}Fe^{2+}Ca(P,Si)Si_6B_3(O,F)_{48}$.

"Garnet" Supergroup

- Nesosilicate insular SiO_4 groups only with cations in and > coordination.
- Category Garnet group, $X_3Z_2(TO_4)_3$ (X = Ca, Fe, etc., Z = Al, Cr, etc., T = Si, As, V, etc.):
  - Pyralspite series:
    - Pyrope $Mg_3Al_2(SiO_4)_3$, Almandine $(Fe^{2+})_3Al_2(SiO_4)_3$, Spessartine $(Mn^{2+})_3Al_2(SiO_4)_3$, Knorringite $Mg_3Cr_2(SiO_4)_3$, Majorite $Mg_3(Fe,Al,Si)_2(SiO_4)_3$, Calderite $(Mn^{2+},-Ca)_3(Fe^{3+},Al)_2(SiO_4)_3$.
  - Ugrandite series:
    - Andradite $Ca_3(Fe^{3+})_2(SiO_4)_3$, Grossular $Ca_3Al_2(SiO_4)_3$, Uvarovite $Ca_3Cr_2(SiO_4)_3$, Goldmanite $Ca_3(V,Al,Fe^{3+})_2(SiO_4)_3$, Yamatoite? $(Mn^{2+},Ca)_3(V^{3+},Al)_2(SiO_4)_3$.
  - Schorlomite - Kimzeyite series:
    - Schorlomite $Ca_3(Ti,Fe^{3+},Al)_2[(Si,Fe^{3+},Fe^{2+})O_4]_3$, Kimzeyite $Ca_3(Z\text{-}r,Ti)_2(Si,Al,Fe^{3+})_3\text{-}O_{12}$, Morimotoite $Ca_3TiFe^{2+}Si_3O_{12}$.
  - Hydrogarnet:
    - Hibschite $Ca_3Al_2(SiO_4)_{3-x}(OH)_{4x}$ (x=0.2-1.5), Katoite $Ca_3Al_2(SiO_4)_{3-x}(OH)_{4x}$ (x=1.5-3).
  - Tetragonal Hydrogarnet:
    - Henritermierite $Ca_3(Mn,Al)_2(SiO_4)_2(OH)_4$, Holtstamite $Ca_3(Al,Mn^{3+})_2(-SiO_4)_2(OH)_4$.
- Bredigite $Ca_7Mg(SiO_4)_4$, Merwinite $Ca_3Mg(SiO_4)_2$, Wadalite $Ca_6Al_5Si_2O_{16}Cl_3$, Rondorfite $Ca_8Mg(SiO_4)4Cl_2$.

"Humite" Supergroup

- Nesosilicate insular SiO_4 groups and O, OH, F, and H_2O with cations in coordination only.
- Topaz group:
  - Topaz $Al_2SiO_4(F,OH)_2$, Krieselite $(Al,Ga)_2(Ge,C)O_4(OH)_2$.
- Humite, general formula $A_n(SiO_4)_m(F,OH)_2$.

- Chondrodite series:
    - Alleghanyite $(Mn^{2+})_5(SiO_4)_2(OH,F)_2$, Chondrodite $(Mg,Fe,Ti)_5(SiO_4)_2(F,OH,O)_2$, Reinhardbraunsite $Ca_5[(OH,F)_2|(SiO_4)_2$, Ribbeite $Mn_5(SiO_4)_2(OH)_2$, Kumtyubeite $Ca_5(SiO_4)_2F_2$.
- Humite series:
    - Humite $(Mg,Fe)_7[(F,OH)_2|(SiO_4)_3]$, Leucophoenicite $(Mn,Ca,Mg,Zn)(SiO_4)_3(OH)_2$, Manganhumite $(Mn^{2+},Mg)_7[(OH)_2|(SiO_4)_3$, Chegemite $Ca_7(SiO_4)_3(OH)_2$.
- Clinohumite series:
    - Clinohumite $(Mg,Fe^{2+})_9[(F,OH)_2|(SiO_4)_4]$, Jerrygibbsite $(Mn,Zn)_9(SiO_4)_4(OH)_2$, Sonolite $Mn_9(SiO_4)_4(F,OH)_2$, Hydroxylclinohumite $Mg_9[(F,OH)_2|(SiO_4)_4$.
- Norbergite $Mg_3(SiO_4)(F,OH)_2$.

- Chloritoid group:
    - Chloritoid $(Fe^{2+},Mg,Mn)_2Al_4Si_2O_{10}(OH)_4$, Magnesiochloritoid $MgAl_2SiO_5(OH)_2$, Ottrelite $(Mn,Fe^{2+},Mg)_2Al_4Si_2O_{10}(OH)_4$, Carboirite-VIII $Fe^{2+}(Al,Ge)_2O[(Ge,Si)O_4](OH)_2$.

Subclass '9.B': Sorosilicates

- Epidote supergroup, $\{A_2\}\{M_3\}[O|OH|SiO_4|Si_2O_7]$:
    - Epidote group:
        - Clinozoisite $\{Ca_2\}\{Al_3\}[O|OH|SiO_4|Si_2O_7]$, Clinozoisite-(Sr) $CaSrAl_3(Si_2O_7)(SiO_4)O(OH)$, Epidote $\{Ca_2\}\{Al_2Fe^{3+}\}[O|OH|SiO_4|Si_2O_7]$, Epidote-(Pb) $CaPbFe^{3+}Al_2(Si_2O_7)(SiO_4)O(OH)$, Mukhinite $\{Ca_2\}\{Al_2V^{3+}\}[O|OH|SiO_4|Si_2O_7]$, Piemontite $\{Ca_2\}\{Al_2Mn^{3+}\}[O|OH|SiO_4|Si_2O_7]$, Piemontite-(Sr) $\{CaSr\}\{Al_2Mn^{3+}\}[O|OH|SiO_4|Si_2O_7]$, Manganipiemontite-(Sr) $\{CaSr\}\{Mn^{3+}AlMn^{3+}\}[O|OH|SiO_4|Si_2O_7]$.
    - Allanite group:
        - Allanite-(Ce) $\{CaCe\}\{Al_2Fe^{2+}\}[O|OH|SiO_4|Si_2O_7]$, Allanite-(La) $\{CaLa\}\{Al_2Fe^{2+}\}[O|OH|SiO_4|Si_2O_7]$, Allanite-(Y) $\{CaY\}\{Al_2Fe^{2+}\}[O|OH|SiO_4|Si_2O_7]$, Dissakisite-(Ce) $\{CaCe\}\{Al_2Mg\}[O|OH|SiO_4|Si_2O_7]$, Dissakisite-(La) $\{CaLa\}\{Al_2Mg\}[O|OH|SiO_4|Si_2O_7]$, Ferriallanite-(Ce) $\{CaCe\}\{Fe^{3+}AlFe^{2+}\}[O|OH|SiO_4|Si_2O_7]$, Manganiandrosite-(Ce) $(Mn^{2+},Ca)(Ce,REE)Mn^{3+}AlMn^{2+}(Si_2O_7)(SiO_4)O(OH)$, Manganiandrosite-(La) $\{LaMn^{2+}\}\{Mn^{3+}AlMn^{2+}\}[O|OH|SiO_4|Si_2O_7]$, Vanadoandrosite-(Ce) $\{Mn^{2+}Ce\}\{V^{3+}AlMn^{2+}\}[O|OH|SiO_4|Si_2O_7]$.
    - Dollaseite group:
        - Dollaseite-(Ce) $\{CaCe\}\{MgAlMg\}[F|OH|SiO_4|Si_2O_7]$, Khristovite-(Ce) $\{CaCe\}\{MgAlMn^{2+}\}[F|OH|SiO_4|Si_2O_7]$, Mills et al. (2009).

Subclass '9.C': Cyclosilicates

- Tourmaline group:

- Alkali-Deficient Tourmaline subgroup - Foitite subgroup:
    - Foitite [ ][$(Fe^{2+})_2(Al,Fe^{3+})$][Al_6][$(OH)_3$|OH|$(BO_3)_3$|Si_6O_{18}], Magnesiofoitite [ ][$Mg_2(Al,Fe^{3+})$][Al_6][$(OH)_3$|OH|$(BO_3)_3$|Si_6O_{18}], Rossmanite [ ][$LiAl_2$][Al_6] [$(OH)_3$|OH|$(BO_3)_3$|Si_6O_{18}], Oxy-Rossmanite [ ][$LiAl_2$]$Al_6(OH)_3O(BO_3)_3$[Si_6O_{18}].
- Calcic Tourmaline subgroup - Liddicoatite subgroup:
    - Liddicoatite [Ca][Li_2Al][Al_6][$(OH)_3$|F|$(BO_3)_3$|Si_6O_{18}], Uvite $CaMg_3(Al_5Mg)(Si_6O_{18})(BO_3)_3(OH)_4$, Feruvite [Ca][$(Fe^{2+},Mg)_3$][$MgAl_5$][$(OH)_3$|F|$(BO_3)_3$|$Si_6O_{18}$], Hydroxy-uvite (IMA2000-030 was not approved, but suspended) $CaMg_3(Al_5Mg)(Si_6O_{18})(BO_3)_3(OH)_3(OH)$.
- Ferric Tourmaline subgroup - Buergerite subgroup:
    - Buergerite [Na][$(Fe^{3+})_3$][Al_6][O_3|F|$(BO_3)_3$|Si_6O_{18}], Povondraite [Na][$(Fe^{3+})_3$] [$(Fe^{3+})_4Mg_2$][$(OH)_3$|O|$(BO_3)_3$|Si_6O_{18}].
- Lithian Tourmaline subgroup - Elbaite subgroup:
    - Olenite $NaAl_9B_3Si_6O_{27}O_3OH$, Oxy-Dravite $Na(MgAl_2)(MgAl_5)Si_6O_{18}(BO_3)_3(OH)_3O$, Elbaite $Na(Al_{1.5}Li_{1.5})Al_6(OH)_3(OH)(BO_3)_3Si_6O_{18}$.
- Sodic Tourmaline subgroup - Schorl subgroup:
    - Dravite [Na][Mg_3][Al_6][$(OH)_3$|OH|$(BO_3)_3$|Si_6O_{18}], Fluor-Dravite $NaMg_3Al_6(OH)_3F(BO_3)_3(Si_6O_{18})$, Schorl [Na][$(Fe^{2+})_3$][$Al_6$][$(OH)_3$|OH|$(BO_3)_3$|-$Si_6O_{18}$], Schorl-(F) [Na][$(Fe^{2+})_3$][$Al_6$][$(OH)_3$|F,OH|$(BO_3)_3$|$Si_6O_{18}$], Chromdravite [Na][$Mg_3$][$(Cr^{3+},Fe^{3+})_6$][$(OH)_3$|OH|$(BO_3)_3$|$Si_6O_{18}$], Vanadiumdravite [Na][$Mg_3$] [$(V^{3+})_6$][$(OH)_3$|OH|$(BO_3)_3$|$Si_6O_{18}$].

- Eudialyte group:
    - Carbokentbrooksite $(Na,[\ ])_{12}(Na,Ce)_3Ca_6Mn_3Zr_3Nb(Si_{25}O_{73})(OH)_3(CO_3)\cdot H_2O$.
    - Eudialyte $Na_4(Ca,Ce)_2(Fe^{2+},Mn,Y)ZrSi_8O_{22}(OH,Cl)_2$ (?).
    - Feklichevite $Na_{11}Ca_9(Fe^{3+},Fe^{2+})_2Zr_3Nb[Si_{25}O_{73}](OH,H_2O,Cl,O)_5$.
    - Ferrokentbrooksite $Na_{15}Ca_6(Fe,Mn)_3Zr_3NbSi_{25}O_{73}(O,OH,H_2O)_3(Cl,F,OH)_2$.
    - Georgbarsanovite $Na_{12}(Mn,Sr,REE)_3Ca_6(Fe^{2+})_3Zr_3NbSi_{25}O_{76}Cl_2\cdot H_2O$.
    - Golyshevite $(Na,Ca)_{10}Ca_9(Fe^{3+},Fe^{2+})_2Zr_3NbSi_{25}O_{72}(CO_3)(OH)_3\cdot H_2O$.
    - Ikranite $(Na,H_3O)_{15}(Ca,Mn,REE)_6(Fe^{3+})_2Zr_3([\ ],Zr)([\ ],Si)Si_{24}O_{66}(O,OH)_6Cl\cdot 2\text{-}3H_2O$.
    - Johnsenite-(Ce) $Na_{12}(Ce,REE,Sr)_3Ca_6Mn_3Zr_3W(Si_{25}O_{73})(CO_3)(OH,Cl)_2$.
    - Kentbrooksite $(Na,REE)_{15}(Ca,REE)_6Mn^{2+}Zr_3NbSi_{25}O_{74}F_2\cdot 2H_2O$.
    - Khomyakovite $Na_{12}Sr_3Ca_6Fe_3Zr_3W(Si_{25}O_{73})(O,OH,H_2O)_3(OH,Cl)_2$.
    - Manganokhomyakovite $Na_{12}Sr_3Ca_6Mn_3Zr_3W(Si_{25}O_{73})(O,OH,H_2O)_3(OH,Cl)_2$.
    - Mogovidite $Na_9(Ca,Na)_6Ca_6Fe_2Zr_3[\ ]Si_{25}O_{72}(CO_3)(OH)_4$.

- Oneillite $Na_{15}Ca_3Mn_3(Fe^{2+})_3Zr_3Nb(Si_{25}O_{73})(O,OH,H_2O)_3(OH,Cl)_2$.
- Raslakite $Na_{15}Ca_3Fe_3(Na,Zr)_3Zr_3(Si,Nb)(Si_{25}O_{73})(OH,H_2O)_3(Cl,OH)$.
- Rastsvetaevite $Na_{27}K_8Ca_{12}Fe_3Zr_6Si_{52}O_{144}(O,OH,H_2O)_6Cl_2$.
- Taseqite $Na_{12}Sr_3Ca_6Fe_3Zr_3NbSi_{25}O_{73}(O,OH,H_2O)_3Cl_2$.
- Zirsilite-(Ce) $(Na,[\ ])_{12}(Ce, Na)_3Ca_6Mn_3Zr_3Nb(Si_{25}O_{73})(OH)_3(CO_3)\cdot H_2O$.
- Alluaivite $Na_{19}(Ca,Mn^{2+})_6(Ti,Nb)_3(Si_3O_9)_2(Si_{10}O_{28})_2Cl\cdot 2H_2O$.
- Andrianovite $Na_{12}(K,Sr,Ce)_3Ca_6Mn_3Zr_3Nb(Si_{25}O_{73})(O,H_2O,OH)_5$.
- Aqualite $(H_3O)_8(Na,K,Sr)_5Ca_6Zr_3Si_{26}O_{66}(OH)_9Cl$.
- Dualite $Na_{30}(Ca,Na,Ce,Sr)_{12}(Na,Mn,Fe,Ti)_6Zr_3Ti_3MnSi_{51}O_{144}(OH,H_2O,Cl)_9$.
- Labyrinthite $(Na,K,Sr)_{35}Ca_{12}Fe_3Zr_6TiSi_{51}O_{144}(O,OH,H_2O)_9Cl_3$, Mills et al. (2009).

Subclass '9.D': Inosilicates

Single Chain Inosilicates

- Astrophyllite group:
    - Astrophyllite $K_2Na(Fe^{2+},Mn)_7Ti_2Si_8O_{26}(OH)_4$.
    - Magnesioastrophyllite $K_2Na[Na(Fe^{2+},Fe^{3+},Mn)Mg_2]Ti_2Si_8O_{26}(OH)_4F$.
    - Hydroastrophyllite $(H_3O,K)_2Ca(Fe^{3+},Mn)_{5\text{-}6}Ti_2Si_8O_{26}(OH)_4F$.
    - Niobophyllite $K_2Na(Fe^{2+},Mn)_7(Nb,Ti)_2Si_8O_{26}(OH)_4(F,O)$.
    - Zircophyllite $K_2(Na,Ca)(Mn,Fe^{2+})_7(Zr,Nb)_2Si_8O_{26}(OH)_4F$.
    - Kupletskite $K_2Na(Mn,Fe^{2+})_7(Ti,Nb)_2Si_8O_{26}(OH)_4F$.
    - Kupletskite-(Cs) $(Cs,K)_2Na(Mn,Fe^{2+},Li)_7(Ti,Nb)_2Si_8O_{26}(OH)_4F$.
    - Niobokupletskite $K_2Na(Mn,Zn,Fe)_7(Nb,Zr,Ti)_2Si_8O_{26}(OH)_4(O,F)$, Mills et al. (2009).
- Sapphirine supergroup:
    - Sapphirine group:
        - Khmaralite $(Mg,Al,Fe)_{16}(Al,Si,Be)_{12}O_{40}$, Sapphirine $(Mg,Al)_8(Al,Si)_6O_{20}$.
    - Aenigmatite group:
        - Aenigmatite $(Na,Ca)_4(Fe^{2+},Ti,Mg)_{12}Si_{12}O_{40}$, Krinovite $NaMg_2CrSi_3O_{10}$, Wilkinsonite $Na_2(Fe^{2+})_4(Fe^{3+})_2Si_6O_{20}$.
    - Rhoenite group:
        - Dorrite $Ca_2Mg_2(Fe^{3+})_4(Al,Fe^{3+})_4Si_2O_{20}$, Hogtuvaite (Ca,Na)2$(Fe^{2+},Fe^{3+},Ti,Mg,Mn)_6(Si,Be,Al)_6O_{20}$, Makarochkinite $Ca_2(Fe^{2+})_4Fe^{3+}TiSi_4BeAlO_{20}$, Rhonite Ca_2(Mg,Fe^{2+},

Fe^{3+},Ti)$_6$(Si,Al)$_6$O$_{20}$, Serendibite $Ca_2(Mg,Al)_6(Si,Al,B)_6O_{20}$, Welshite $Ca_4Mg_9Sb_3O_4[Si_6Be_3AlFe_2O_{36}]$.

- Surinamite $(Mg,Fe^{2+})_3Al_4BeSi_3O_{16}$, Mills et al. (2009).

Pyroxene Supergroup

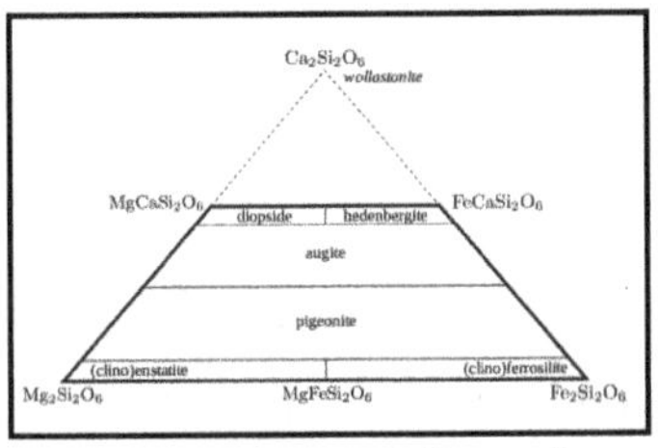

Pyroxene Quadrilateral

- Orthopyroxene group:
  - Donpeacorite $(Mn^{2+},Mg)Mg[SiO_3]_2$, Enstatite $MgSiO_3$, Ferrosilite, $FeSiO_3$.
- Clinopyroxene group:
  - Aegirine $NaFe^{3+}Si_2O_6$, Augite $(Ca,Na)(Mg,Fe^{2+},Al,Fe^{3+},Ti)[(Si,Al)_2O_6]$, Clinoenstatite $MgSiO_3$, Clinoferrosilite $Fe^{2+}SiO_3$, Diopside $CaMg[Si_2O_6]$, Esseneite $CaFe^{3+}[AlSiO_6]$, Grossmanite $CaTi^{3+}AlSiO_6$, Hedenbergite $CaFe^{2+}[Si_2O_6]$, Jadeite $Na(Al,Fe^{3+})[Si_2O_6]$, Jervisite $(Na,Ca,Fe^{2+})(Sc,Mg,Fe^{2+})[Si_2O_6]$, Johannsenite $CaMn^{2+}[Si_2O_6]$, Kanoite $Mn^{2+}(Mg,Mn^{2+})[Si_2O_6]$, Kosmochlor $NaCr[Si_2O_6]$, Kushiroite $CaAl[Si_2O_6]$, Namansilite NaMn3+$[Si_2O_6]$, Natalyite $Na(V^{3+},Cr)[Si_2O_6]$, Petedunnite $Ca(Zn,Mn^{2+},Mg,Fe^{2+})[Si_2O_6]$, Pigeonite $(Mg,Fe^{2+},Ca)(Mg,Fe^{2+})Si_2O_6$, Spodumene $LiAlSi_2O_6$.

Multiple Chain Inosilicates

Note: the amphibole subcommittee (CNMNC/ IMA) published many reports (IMA 1978 s.p., IMA 1997 s.p., IMA 2003 s.p., IMA 2012 s.p.), renaming and redefining many minerals. Working draft: rruff.info, mindat.org and mineralienatlas.de are not up to date yet.

Amphibole Supergroup

- W(OH, F, Cl)-dominant amphibole: calcic subgroup:
  - Cannilloite root name: fluoro-cannilloite $CaCa_2Mg_4Al(Si_5Al_3)O_{22}(OH)_2$ (1993-033, IMA 1997 s.p. Rd, IMA 2012 s.p. Rd Rn from cannilloite).
  - Edenite root name: edenite $Na[Ca_2][Mg_5][(OH)_2|AlSi_7O_{22}]$ (1839, IMA 2012 s.p. Rd), ferro-edenite $[Na][Ca_2][(Fe^{2+})_5][(OH)_2|AlSi_7O_{22}]$ (1946, IMA 1997 s.p. Rd, IMA 2012 s.p. Rd), fluoro-edenite $Na[Ca_2][Mg_5][(F,OH)_2|AlSi_7O_{22}]$ (IMA 1994-059, IMA 2012 s.p. Rd).
  - Hastingsite root name: hastingsite $[Na][Ca_2][(Fe^{2+})_4Fe^{3+}][(OH)_2|Al_2Si_6O_{22}]$ (1896, IMA 2012 s.p. Rd), magnesio-fluoro-hastingsite $(Na,K)Ca_2(Mg,Fe^{3+},Ti)_5(Si,Al)_8O_{22}F_2$ (IMA 2005-002, IMA 2012 s.p. Rd Rn from fluoro-magnesiohastingsite), magnesio-hastingsite $Na[Ca_2][Mg_4Fe^{3+}][(OH)_2|Al_2Si_6O_{22}]$ (1928, IMA 1997 s.p. Rd, IMA 2012 s.p. Rd

Rn from magnesiohastingsite), potassic-fluoro-hastingsite KCa2($(Fe^{2+})_2$,Mg_2,Fe^{3+})S_5(-Si_6Al_2)$_8O_{22}F_2$ (IMA 2005-006, IMA 2012 s.p. Rd Rn from fluoro-potassichastingsite), potassic-chloro-hastingsite $KCa_2(Mg_4Al)(Si_6Al_2)O_{22}Cl_2$ (IMA 2005-007, IMA 2012 s.p. Rd Rn from chloro-potassicpargasite, syn. dashkesanite), potassic-magnesio-hastingsite $(K,Na)Ca_2(Mg,Fe^{2+},Fe^{3+},Al)_5(Si,Al)_8O_{22}(OH,Cl)_2$ (IMA 2004-027b, IMA 2012 s.p. Rd Rn from potassic-magnesiohastingsite).

- Joesmithite $PbCa_2(Mg,Fe^{2+},Fe^{3+})_5Si_6Be_2O_{22}(OH)_2$ (1968, IMA 2012 s.p. Rd).
- Magnesio-hornblende root name: ferro-hornblende $[Ca_2][(Fe^{2+})_4Al][(OH)_2|AlSi_7O_{22}]$ (1930, IMA 1978 s.p., IMA 1997 s.p. Rn from ferro-hornblende, IMA 2012 s.p. Rd Rn from ferrohornblende), magnesio-hornblende $[Ca_2][Mg_4Al][(OH)_2|AlSi_7O_{22}]$ (1965, IMA 1997 s.p. Rd, IMA 2012 s.p. Rd Rn from magnesiohornblende).
- Pargasite root name: chromio-pargasite (IMA 2011-023, IMA 2012 s.p. Rd Rn from ehimeite), ferro-pargasite $[Na][Ca_2][(Fe^{2+})_4Al][(OH)_2|Al_2Si_6O_{22}]$ (1961, IMA 1997 s.p. Rd, IMA 2012 s.p. Rd Rn from ferropargasite), fluoro-pargasite $NaCa_2(Mg_3Fe^{2+}Al)_5$(-$Si_6Al_2O_{22})F_2$ (IMA 2003-050, IMA 2012 s.p. Rd Rn from fluoropargasite), pargasite $[Na][Ca_2][Mg_4Al][(OH)_2|Al_2Si_6O_{22}]$ (1815, IMA 1997 s.p. Rd, IMA 2012 s.p. Rd), potassic-chloro-pargasite $KCa_2((Fe^{2+})_3MgFe^{3+})(Si_6Al_2)S_8O_{22}Cl_2$ (IMA 2001-036, IMA 2012 s.p. Rd Rn from chloro-potassichastingsite), potassic-ferro-pargasite $KCa_2((Fe^{2+})_4Al)Si_6Al_2O_{22}(OH)_2$ (IMA 2007-053, IMA 2012 s.p. Rd Rn from potassic-ferropargasite), potassic-fluoro-pargasite (IMA 2009-091, IMA 2012 s.p. Rd Rn from fluoro-potassic-pargasite), potassic-pargasite $[K][Ca_2][Mg_4Al][(OH)_2|Al_2Si_6O_{22}]$ (IMA 1994-046, IMA 2012 s.p. Rd Rn from potassicpargasite).
- Sadanagaite root name: potassic-ferro-ferri-sadanagaite $[K][Ca_2][(Fe^{2+})_3(Fe^{3+})_2][(OH)_2|Al_3Si_5O_{22}]$ (IMA 1997-035, IMA 2012 s.p. Rd Rn from potassic-ferrisadanagaite), potassic-ferro-sadanagaite $KCa_2Fe^{2+}{}_3(Al,Fe^{3+})_2(Si_5Al_3)O_{22}(OH)_2$ (IMA 1980-027, 2004 Rd, IMA 1997 s.p. Rd Rn from sadanagaite, IMA 2012 s.p. Rd Rn from potassicsadanagaite), potassic-sadanagaite $(K,Na)Ca_2(Mg,Fe^{2+},Al,Ti)_5[(Si,Al)_8O_{22}]$ (IMA 1982-102, 2004 Rd, IMA 2003 s.p. Rn from magnesio-sadanagaite, IMA 2012 s.p. Rd Rn from potassic-magnesiosadanagaite), sadanagaite $[Na][Ca_2][Mg_3Al_2][(OH)_2|Al_3Si_5O_{22}]$ (1984, IMA 1997 s.p. Rd Rn from sadanagaite, IMA 2002-051, 2004 Rn from potassic-magnesiosadanagaite, IMA 2012 s.p. Rd Rn from magnesiosadanagaite).
- Tremolite-actinolite root name: actinolite $Ca_2(Mg,Fe^{2+})_5(Si_8O_{22})(OH)_2$ (1794, IMA 2012 s.p. Rd), ferro-actinolite $[Ca_2][(Fe^{2+},Mg)_5][(OH)_2|Si_8O_{22}]$ (1946, IMA 1997 s.p. Rd, IMA 2012 s.p. Rd), tremolite $[Ca_2][Mg_5][(OH)_2|Si_8O_{22}]$ (1789, IMA 1997 s.p. Rd, IMA 2012 s.p. Rd).
- Tschermakite root name: tschermakite $[Ca_2][Mg_3Fe^{3+}Al][(OH)_2|Al_2Si_6O_{22}]$ (1945, IMA 1997 s.p. Rd, IMA 2012 s.p. Rd).

- W(OH, F, Cl)-dominant amphibole: lithium subgroup:
  - Clino-holmquistite root name: clino-ferro-ferri-holmquistite $Li_2((Fe^{3+})_2(Fe^{2+})_3)Si_8O_{22}(OH)_2$ (IMA 1997 s.p., 2001-066, IMA 2012 s.p. Rd Rn from ferri-clinoferro-holmquistite).

- Holmquistite root name: ferro-holmquistite $(Li_2(Fe^{2+})_3Al_2)Si_8O_{22}(OH)_2$ (IMA 2004-030, IMA 2012 s.p. Rd), holmquistite $(Li_2Mg_3Al_2)Si_8O_{22}(OH)_2$ (1913, IMA 1997 s.p. Rd, IMA 2012 s.p. Rd).
- Pedrizite root name: ferri-pedrizite $NaLi_2((Fe^{3+})_2Mg_2Li)Si_8O_{22}(OH)_2$ (IMA 2001-032, IMA 2003 s.p. discredited, IMA 2012 s.p. Rd revalidated), ferro-ferri-pedrizite $NaLi_2((Fe^{3+})_2(Fe^{2+})_3)Si_8O_{22}(OH)_2$ (IMA 2003 s.p., IMA 2012 s.p. Rd Rn from sodic-ferro-ferripedrizite), ferro-fluoro-pedrizite $NaLi_2(Fe^{2+})_2Al_2Li)Si_8O_{22}F_2$ (IMA 2008-070, IMA 2012 s.p. Rd Rn from fluoro-sodic-ferropedrizite), fluoro-pedrizite $NaLi_2(Mg_2Al_2\text{-}Li)Si_8O_{22}F_2$ (IMA 2004-002, IMA 2012 s.p. Rd Rn from fluoro-sodic-pedrizite).

- W(OH, F, Cl)-dominant amphibole: Mg-Fe-Mn subgroup:
  - Anthophyllite root name: anthophyllite $Mg_7Si_8O_{22}(OH)_2$ (1801, IMA 2012 s.p. Rd), ferro-anthophyllite $(Fe^{2+})_7Si_8O_{22}(OH)_2$ (1821, IMA 1997 s.p. Rd, IMA 2012 s.p. Rd), proto-anthophyllite $(Mg,Fe)_7Si_8O_{22}(OH)_2$ (IMA 2001-065, IMA 2012 s.p. Rd), proto-ferro-anthophyllite $(Fe^{2+},Mn^{2+})_2(Fe^{2+},Mg)_5(Si_4O_{11})_2(OH)_2$ (IMA 1986-006, IMA 2012 s.p. Rd), protomangano-ferro-anthophyllite $(Mn^{2}+,Fe^{2}+)_2(Fe^{2+},Mg)_5(Si_4O_{11})_2(OH)_2$ (IMA 1986-007, IMA 2012 s.p. Rd).
  - Gedrite root name: ferro-gedrite $(Fe^{2+})_5Al_2Si_6Al_2O_{22}(OH)_2$ (1939, IMA 1978 s.p. Rn, IMA 1997 s.p. Rd, IMA 2012 s.p. Rd Rn from ferrogedrite), gedrite $Mg_5Al_2Si_6Al_2O_{22}(OH)_2$ (1836, IMA 2012 s.p. Rd).
  - Minerals: cummingtonite $Mg_7Si_8O_{22}(OH)_2$ (1824, IMA 1997 s.p. Rd, IMA 2012 s.p. Rd), grunerite $(Fe^{2+})_7Si_8O_{22}(OH)_2$ (1853, IMA 1997 s.p. Rd, IMA 2012 s.p. Rd).
- W(OH, F, Cl)-dominant amphibole: sodic subgroup:
  - Arfvedsonite root name: arfvedsonite $NaNa_2((Fe^{2+})_4Fe^{3+})Si_8O_{22}(OH)_2$ (1823, IMA 2012 s.p. Rd), magnesio-arfvedsonite $NaNa_2(Mg_4Fe^{2+})Si_8O_{22}(OH)_2$ (1957, IMA 2012 s.p. Rd), magnesio-fluoro-arfvedsonite $NaNa_2(Mg,Fe^{2+})_4Fe^{3+}[Si_8O_{22}](F,OH)_2$ (IMA 1998-056, IMA 2012 s.p. Rd Rn from fluoro-magnesio-arfvedsonite), potassic-arfvedsonite $KNa_2(Fe^{2+})_4Fe^{3+}Si_8O_{22}(OH)_2$ (IMA 2003-043, IMA 2012 s.p. Rd), potassic-magnesio-fluoro-arfvedsonite $KNa_2(Mg_4Fe^{3+})Si_8O_{22}F_2$ (IMA 1985-023, 2006 Rn from potassium fluor-magnesio-arfvedsonite, 2010 Rd, IMA 2012 s.p. Rd Rn from fluoro-potassic-magnesio-arfvedsonite).
  - Eckermannite root name: eckermannite $NaNa_2(Mg_4Al)Si_8O_{22}(OH)_2$ (1942, IMA 2012 s.p. Rd), mangano-ferri-eckermannite $NaNa_2(Mn^{2+})_4(Fe^{3+},Al)Si_8O_{22}(OH)_2$ (1968-028, IMA 1997 s.p., IMA 2012 s.p. Rd Rn from kôzulite).
  - Glaucophane root name: ferro-glaucophane $Na_2((Fe^{2+})_3Al_2)Si_8O_{22}(OH)_2$ (1957, IMA 1997 s.p. Rd, IMA 2012 s.p. Rd Rn from ferroglaucophane), glaucophane $Na_2(Mg_3Al_2)Si_8O_{22}(OH)_2$ (1963, IMA 1997 s.p. Rd, IMA 2012 s.p. Rd).
  - Leakeite root name: ferri-fluoro-leakeite $NaNa_2(Mg_2Fe^{3+}{}_2Li)Si_8O_{22}F_2$ (IMA 2009-085, IMA 2012 s.p. Rd Rn from fluoroleakeite), ferri-leakeite $Na(NaLi)(Mg_2(Fe^{3+})_2Li)Si_8O_{22}(OH)_2$ (IMA 2001-069, IMA 2012 s.p. Rd Rn from ferriwhittakerite), ferro-ferri-fluoro-leakeite $NaNa_2((Fe^{2+})_2(Fe^{3+})_2Li)Si_8O_{22}F_2$ (IMA 1993-026, IMA 2012

s.p. Rd Rn from fluoro-ferroleakeite), fluoro-leakeite $NaNa_2(Mg_2Al_2Li)Si_8O_{22}F_2$ (IMA 2009-012, IMA 2012 s.p. Rd Rn from fluoro-aluminoleakeite), potassic-ferri-leakeite $KNa_2Mg_2(Fe^{3+})_2LiSi_8O_{22}(OH)_2$ (IMA 2001-049, IMA 2012 s.p. Rd Rn from potassicleakeite), potassic-leakeite $KNa_2(Mg_2Al_2Li)Si_8O_{22}(OH)_2$ (2002, IMA 2012 s.p. Rd), potassic-mangani-leakeite $KNa_2(Mg_2Mn^{3+}{}_2Li)Si_8O_{22}(OH)_2$ (IMA 1992-032, IMA 2012 s.p. Rd Rn from kornite).

- Nybøite root name: ferro-ferri-nybøite $NaNa_2(Fe^{2+})_3(Fe^{3+})_2(Si_7Al)O_{22}(OH)_2$ (IMA 1997 s.p., IMA 2012 s.p. Rd Rn from ferric-ferronyboite), fluoro-nybøite $NaNa_2(Al_2Mg_3)(Si_7Al)O_{22}(F,OH)_2$ (IMA 2002-010, IMA 2012 s.p. Rd Rn from fluoronybøite), nybøite $NaNa_2(Mg_3Al_2)Si_7AlO_{22}(OH)_2$ (1981, IMA 1997 s.p. Rd, IMA 2012 s.p. Rd).
- Riebeckite root name: fluoro-riebeckite $Na_2(Fe^{2+}{}_3Fe^{3+}{}_2)Si_8O_{22}F_2$ (1966, IMA 2012 s.p. Rd), magnesio-riebeckite $Na_2[(Mg,Fe^{2+})_3(Fe^{3+})_2]Si_8O_{22}(OH)_2$ (1957, IMA 1997 s.p. Rd, IMA 2012 s.p. Rd), riebeckite $Na_2((Fe^{2+})_3(Fe^{3+})_2)Si_8O_{22}(OH)_2$ (1888, IMA 1997 s.p. Rd, IMA 2012 s.p. Rd).

- W(OH, F, Cl)-dominant amphibole: sodic-calcic subgroup:
  - Barroisite root name: barroisite $[CaNa][Mg_3AlFe^{3+}][(OH)_2|AlSi_7O_{22}]$ (1922, IMA 1997 s.p. Rd, IMA 2012 s.p. Rd).
  - Katophorite root name: ferri-fluoro-katophorite $Na(NaCa)(Mg_4Fe^{3+})(Si_7Al)O_{22}F_2$ (IMA 2012 s.p.), ferri-katophorite $Na_2Ca(Fe^{2+},Mg)_4Fe^{3+}(Si_7Al)O_{22}(OH)_2$ (IMA 1978 s.p., IMA 2012 s.p. Rd Rn from ferrikatophorite), ferro-katophorite $Na(NaCa)(Fe^{2+}{}_4Al)(Si_7Al)O_{22}(OH)_2$ (1894, IMA 2012 s.p. Rd Rn), katophorite $Na(NaCa)(Mg_4Al)(Si_7Al)O_{22}(OH)_2$ (1894, IMA 1997 s.p. Rd, IMA 2012 s.p. Rd).
  - Richterite root name: ferro-richterite $Na[CaNa][(Fe^{2+})_5][(OH)_2|Si_8O_{22}]$ (1946, IMA 1997 s.p. Rd, IMA 2012 s.p. Rd Rn from ferrorichterite), fluoro-richterite $Na(CaNa)Mg_5[Si_8O_{22}]F_2$ (IMA 1992-020, IMA 2012 s.p. Rd Rn from fluororichterite), potassic-fluoro-richterite $K[CaNa][Mg_5][(F,OH)_2|Si_8O_{22}]$ (IMA 1986-046, IMA 2004 s.p. Rn from potassium-fluorrichterite, IMA 2012 s.p. Rd Rn from fluoro-potassicrichterite), richterite $Na[CaNa][Mg_5][(OH)_2|Si_8O_{22}]$ (1865, IMA 1997 s.p. Rd, IMA 2012 s.p. Rd).
  - Taramite root name: ferro-taramite $Na[CaNa][(Fe^{2+})_3Al_2][(OH)_2|Al_2Si_6O_{22}]$ (IMA 2006-023, IMA 2012 s.p. Rd Rn from aluminotaramite), fluoro-taramite $Na_2CaMg_3Al_2(Si_6Al_2)O_{22}F_2$ (IMA 2006-025, IMA 2012 s.p. Rd Rn from fluoro-alumino-magnesiotaramite), potassic-ferro-ferri-taramite $K(CaNa)(Fe^{2+}{}_3Fe^{3+}{}_2)(Si_6Al_2)O_{22}(OH)_2$ (IMA 1964-003, IMA 1978 s.p. Rn from mboziite, IMA 1997 s.p. Rn from ferri-taramite, IMA 2012 s.p. Rd Rn from ferritaramite, Erratum 2013 Rd Rn from ferro-ferri-taramite), potassic-ferro-taramite $K(CaNa)(Fe^{2+}{}_3Al_2)(Si_6Al_2)O_{22}(OH)_2$ (IMA 2007-015, IMA 2012 s.p. Rd Rn from potassic-aluminotaramite), taramite $Na_2CaMg_3Al_2(Si_6Al_2)O_{22}(OH)_2$ (IMA 2006-024, IMA 2012 s.p. Rd Rn from alumino-magnesiotaramite).
  - Winchite root name: ferri-winchite $(Na)(Na,Ca)_2(Mg,Fe^{2+})_4Fe^{3+}[Si_8O_{22}](OH)_2$ (IMA 2004-034, IMA 2012 s.p. Rd Rn from ferriwinchite), ferro-winchite $[CaNa][(Fe^{2+})_4(Al,Fe^{3+})][(OH)_2|Si_8O_{22}]$ (IMA 1978 s.p., IMA 1997 s.p. Rd, IMA 2012 s.p. Rd

Rn from ferrowinchite), winchite $[CaNa][Mg_4(Al,Fe^{3+})][(OH)_2|Si_8O_{22}]$ (1906, IMA 1997 s.p. Rd, IMA 2012 s.p. Rd).

- W(OH, F, Cl)-dominant amphibole: Na-Mg-Fe-Mn subgroup:
  - Ghoseite root name: ferri-ghoseite $Na(NaMn^{2+})(Mg_4Fe^{3+})Si_8O_{22}(OH)_2$ (IMA 2003-066, IMA 2012 s.p. Rd Rn from parvowinchite).
- W(O)-dominant amphibole group:
  - Kaersutite root name: ferri-kaersutite $NaCa_2(Mg_3Fe^{3+}Ti)(Si_6Al_2)O_{22}O_2$ (IMA 2011-035, IMA 2012 s.p. Rd Rn from ferrikaersutite), kaersutite $NaCa_2(Mg_4Ti)(Si_6Al_2)O_{23}(OH)$ (IMA 1997 s.p., IMA 2012 s.p. Rd).
  - Minerals: mangani-dellaventuraite $NaNa_2(Mg_2,Mn^{3+},Li,Ti)Si_8O_{22}O_2$ (IMA 2003-061, IMA 2012 s.p. Rd Rn from dellaventuraite), mangano-mangani-ungarettiite $NaNa_2((Mn^{2+})_2(Mn^{3+})_3)Si_8O_{22}O_2$ (IMA 1994-004, IMA 2012 s.p. Rd Rn from ungarettiite).

Subclass '9.E': Phyllosilicates

- Category:Clay minerals group.
- Category:Medicinal clay.
- Category:Mica group.
- Category:Serpentine group/ Kaolinite-Serpentine group.
- Category:Smectite group/ Montmorillonite group.
- Pyrophyllite-Talc group.
- Chlorite group: Clinochlore, Nimite, Pennantite, Baileychlore, Cookeite, Donbassite, Gonyerite, Odinite, Sudoite, Orthochamosite.

Subclass '9.F': Tectosilicates

Tectosilicates without zeolitic H_2O

- Category:Quartz varieties.
- Feldspar family.
- Feldspathoid family:
  - Sodalite group.
  - Helvine group.
  - Category:Cancrinite group.

Tectosilicates with Zeolitic H_2O

- Category:Zeolite group.

- Alflarsenite $NaCa_2Be_3Si_4O_{13}(OH){\bullet}2H_2O$.
- Zeolites with T_5O_{10} Units – The Fibrous Zeolites
    - Natrolite subgroup:
        - Gonnardite $(Na,Ca)_2(Si,Al)_5O_{10}{\cdot}3H_2O$, Mesolite $Na_2Ca_2Si_9Al_6O_{30}{\cdot}8H_2O$, Natrolite $Na_2Al_2Si_3O_{10}{\cdot}2H_2O$, Paranatrolite $Na_2Al_2Si_3O_{10}{\cdot}3H_2O$, Scolecite $CaAl_2Si_3O_{10}{\cdot}3H_2O$.
    - Tetranatrolite? $Na_2[Al_2Si_3O_{10}]{\bullet}2H_2O$, Thomsonite-Sr $(Sr,Ca)_2Na[Al_5Si_5O_{20}]{\bullet}7H_2O$, Thomsonite-Ca $NaCa_2Al_5Si_5O_{20}{\bullet}6H_2O$, Kalborsite $K6Al_4Si_6BO_{20}(OH)_4Cl$, Edingtonite $BaAl_2Si_3O_{10}{\bullet}4H_2O$.
- Chains of single connected 4-membered rings:
    - Ammonioleucite $(NH_4,K)AlSi_2O_6$, Leucite $KAlSi_2O_6$, Analcime $NaAlSi_2O_6{\bullet}H_2O$, Hsianghualite $Ca_3Li_2Be_3(SiO_4)_3F_2$, Lithosite $K_6Al_4Si_8O_{25}{\bullet}H_2O$, Pollucite $(Cs,Na)_2Al_2Si_4O_{12}{\bullet}H_2O$, Wairakite $CaAl_2Si_4O_{12}{\bullet}2H_2O$, Laumontite $CaAl_2Si_4O_{12}{\bullet}4H_2O$, Yugawaralite $CaAl_2$-$Si_6O_{16}{\bullet}4H_2O$, Roggianite $Ca_2[Be(OH)_2Al_2Si_4O_{13}]{\bullet}2.5H_2O$, Goosecreekite $CaAl_2Si_6O_{16}{\bullet}5H_2O$, Montesommaite $(K,Na)_9Al_9Si_{23}O_{64}{\bullet}10H_2O$, Partheite $Ca_2Al_4Si_4O_{15}(OH)_2{\bullet}4H_2O$.
- Chains of doubly connected 4-membered rings:
    - Amicite $K_2Na_2Al_4Si_4O_{16}{\bullet}5H_2O$, Garronite $Na_2Ca_5Al_{12}Si_{20}O_{64}{\bullet}27H_2O$, Gobbinsite $(Na_2$,-$Ca)_2K_2Al_6Si_{10}O_{32}{\bullet}12H_2O$, Gismondine $Ca_2Al_4Si_4O_{16}{\bullet}9H_2O$, Harmotome $(Ba,Na,K)_{(1-2)}(Si,Al)_8O_{16}{\bullet}6H_2O$, Phillipsite-Na $(Na,K,Ca)_{(1-2)}(Si,Al)_8O_{16}{\bullet}6H_2O$, Phillipsite-Ca $(Ca,K,Na)_{(1-2)}(Si,Al)_8O_{16}{\bullet}6H_2O$, Phillipsite-K $(K,Na,Ca)_{(1-2)}(Si,Al)_8O_{16}{\bullet}6H_2O$, Merlinoite $(K,Ca,Na,Ba)_7Si_{23}Al_9O_{64}{\bullet}23H_2O$, Mazzite-Mg $K_2CaMg_2(Al,Si)_{36}O_{72}{\bullet}28H_2O$, Mazzite-Na $Na_8Al_8Si_{28}O_{72}{\bullet}30H_2O$, Perlialite $K_8Tl_4Al_{12}Si_{24}O_{72}{\bullet}20H_2O$, Boggsite $NaCa_2(Al_5$-$Si_{19}O_{48}){\bullet}17H_2O$, Paulingite-Ca $(Ca,K,Na,Ba)_5[Al_{10}Si_{35}O_{84}]{\bullet}34H_2O$, Paulingite-K $(K_2$,-$Ca,Na_2,Ba)_5Al_{10}Si_{35}O_{90}{\bullet}45H_2O$, Paulingite-Na $(Na_2,K_2,Ca,Ba)_5Al_{10}Si_{35}O_{90}{\bullet}45H_2O$.
- Chains of 6-membered rings – tabular zeolites:
    - Gmelinite-Ca $(Ca,Na_2)Al_2Si_4O_{12}{\bullet}6H_2O$, Gmelinite-K $(K,Na,Ca)_6(Al_7Si_{17}O_{48}){\bullet}22H_2O$, Gmelinite-Na $(Na_2,Ca)Al_2Si_4O_{12}{\bullet}6H_2O$, Chabazite-K $(K_2,Ca,Na_2,Mg)[Al_2Si_4O_{12}]{\bullet}6H_2O$, Chabazite-Ca $(Ca_{0.5},Na,K)_4[Al_4Si_8O_{24}]{\bullet}12H_2O$, Chabazite-Na $(Na_2,K_2,Ca,Mg)[Al_2Si_4O_{12}]{\bullet}6H_2O$, Chabazite-Sr $(Sr,Ca,K_2,Na_2)[Al_2Si_4O_{12}]{\bullet}6H_2O$, Herschelite? $(Na,Ca,K)AlSi_2O_6{\bullet}3H_2O$, Willhendersonite $KCaAl_3Si_3O_{12}{\bullet}5H_2O$, Levyne-Ca $(Ca,Na_2,K_2)Al_2Si_4O_{12}{\bullet}6H_2O$, Levyne-Na $(Na_2$,-$Ca,K_2)Al_2Si_4O_{12}{\bullet}6H_2O$, Bellbergite $(K,Ba,Sr)_2Sr_2Ca_2(Ca,Na)_4Al_{18}Si_{18}O_{72}{\bullet}30H_2O$, Erionite-Ca $(Ca,K_2,Na_2)_2[Al_4Si_{14}O_{36}]{\bullet}15H_2O$, Erionite-K $(K_2,Ca,Na_2)_2[Al_4Si_{14}O_{36}]{\bullet}15H_2O$, Erionite-Na $(Na_2,K_2,Ca)_2[Al_4Si_{14}O_{36}]{\bullet}15H_2O$, Wenkite $Ba_4Ca_6(Si,Al)_{20}O_{39}(OH)_2(SO_4)_3{\bullet}nH_2O$ (?), Offretite $(K_2,Ca,Mg)_{2.5}Al_5Si_{13}O_{36}{\bullet}15H_2O$, Faujasite-Ca $(Ca,Na_2,Mg)_{3.5}[Al_7Si_{17}O_{48}]{\bullet}32H_2O$, Faujasite-Mg $(Mg,Na_2,Ca)_{3.5}[Al_7Si_{17}O_{48}]{\bullet}32H_2O$, Faujasite-Na $(Na_2,Ca,Mg)_{3.5}[Al_7Si_{17}O_{48}]{\bullet}32H_2O$, Maricopaite $Pb_7Ca_2(Si,Al)_{48}O_{100}{\bullet}32H_2O$, Mordenite $(Ca,Na_2,K_2)Al_2Si_{10}O_{24}{\bullet}7H_2O$, Dachiardite-Ca $(Ca,Na_2,K_2)_5Al_{10}Si_{38}O_{96}{\bullet}25H_2O$, Dachiardite-Na $(Na_2,Ca,K_2)_4Al_4Si_{20}O_{48}{\bullet}13H_2O$, Epistilbite $CaAl_2Si_6O_{16}{\bullet}5H_2O$, Ferrierite-K $(K,Na)_2Mg(Si,Al)_{18}O_{36}{\bullet}9H_2O$, Ferrierite-Mg $(Mg,Na,K)_2Mg(Si,Al)_{18}O_{36}{\bullet}9H_2O$, Ferrierite-Na $(Na,K)_2Mg(Si,Al)_{18}O_{36}{\bullet}9H_2O$, Bikitaite $Li_2[Al_2Si_4O_{12}]{\bullet}2H_2O$.
- Chains of $T_{10}O_{20}$ Tetrahedra:

- Clinoptilolite-Na $(Na,K,Ca)_{(2\text{-}3)}Al_3(Al,Si)_2Si_{13}O_{36}\cdot 12H_2O$, Clinoptilolite-K $(Na,K,Ca)_{(2\text{-}3)}Al_3(Al,Si)_2Si_{13}O_{36}\cdot 12H_2O$, Clinoptilolite-Ca $(Ca,Na,K)_{(2\text{-}3)}Al_3(Al,Si)_2Si_{13}O_{36}\cdot 12H_2O$, Heulandite-Ba $(Ba,Ca,K,Na,Sr)_5Al_9Si_{27}O_{72}\cdot 22H_2O$, Heulandite-Ca $(Ca,Na)_{(2\text{-}3)}Al_3(Al,Si)_2Si_{13}O_{36}\cdot 12H_2O$, Heulandite-K $(K,Na,Ca)_{(2\text{-}3)}Al_3(Al,Si)_2Si_{13}O_{36}\cdot 12H_2O$, Heulandite-Na $(Na,Ca)_{(2\text{-}3)}Al_3(Al,Si)_2Si_{13}O_{36}\cdot 12H_2O$, Heulandite-Sr $(Sr,Na,Ca)_{(2\text{-}3)}Al_3(Al,Si)_2Si_{13}O_{36}\cdot 12H_2O$, Stilbite-Ca $NaCa_4[Al_8Si_{28}O_{72}]\cdot nH_2O$ (n=28-32), Stilbite-Na $Na_3Ca_3[Al_8Si_{28}O_{72}]\cdot nH_2O$ (n=28-32), Barrerite $(Na,K,Ca)_2Al_2Si_7O_{18}\cdot 6H_2O$, Stellerite $CaAl_2Si_7O_{18}\cdot 7H_2O$, Brewsterite-Ba $(Ba,Sr)Al_2Si_6O_{16}\cdot 5H_2O$, Brewsterite-Sr $(Sr,Ba)Al_4Si_{12}O_{32}\cdot 10H_2O$.

- Other Rare Zeolites:
  - Terranovaite $(Na,Ca)_8(Si_{68}Al_{12})O_{160}\cdot 29H_2O$, Gottardiite $Na_3Mg_3Ca_5Al_{19}Si_{117}O_{272}\cdot 93H_2O$, Lovdarite $K_2Na_6Be_4Si_{14}O_{36}\cdot 9H_2O$, Gaultite $Na_4Zn_2Si_7O_{18}\cdot 5H_2O$, Chiavennite $CaMnBe_2Si_5O_{13}(OH)_2\cdot 2H_2O$, Tschernichite $(Ca,Na)(Si_6A_{12})O_{16}\cdot(4\text{-}8)H_2O$, Mutinaite $Na_3Ca_4Si_{85}Al_{11}O_{192}\cdot 60H_2O$, Tschortnerite $Ca_4(Ca,Sr,K,Ba)_3Cu_3(OH)_8[Si_{12}Al_{12}O_{48}]\cdot xH_2O$, x>=20, Thornasite $Na_{12}Th_3[Si_8O_{19}]_4\cdot 18H_2O$, Direnzoite $NaK_6MgCa_2(Al_{13}Si_{47}O_{120})\cdot 36H_2O$.

- Unclassified zeolites:
  - Cowlesite $CaAl_2Si_3O_{10}\cdot(5\text{-}6)H_2O$, Mountainite $(Ca,Na_2,K_2)_2Si_4O_{10}\cdot 3H_2O$.

Native Element Minerals

Native copper

Native gold

Native silver

Native sulfur

Diamond

Native element minerals are those elements that occur in nature in uncombined form with a distinct mineral structure. The elemental class includes metals and intermetallic elements, naturally occurring alloys, semi-metals and non-metals. The Nickel–Strunz classification system also includes the naturally occurring phosphides, silicides, nitrides and carbides.

Elements

The following elements occur as native element minerals or alloys (list of minerals (complete)):

- Aluminium,
- Antimony,
- Arsenic,
- Bismuth,
- Carbon,
- Cadmium,
- Chromium,

- Copper,
- Gold,
- Indium,
- Iron,
- Iridium,
- Lead,
- Mercury,
- Nickel,
- Osmium,
- Palladium,
- Platinum,
- Rhenium,
- Rhodium,
- Selenium,
- Silver,
- Silicon,
- Sulfur,
- Tantalum,
- Tellurium,
- Tin,
- Titanium,
- Vanadium,
- Zinc.

Nickel–Strunz Classification -01- Native elements

This list uses the Classification of Nickel–Strunz (mindat.org, 10 ed, pending publication).

- Abbreviations:
    - "*" - discredited (IMA/CNMNC status).
    - "?" - questionable/doubtful (IMA/CNMNC status).
    - "REE" - Rare-earth element (Sc, Y, La, Ce, Pr, Nd, Pm, Sm, Eu, Gd, Tb, Dy, Ho, Er, Tm, Yb, Lu).

- "PGE" - Platinum-group element (Ru, Rh, Pd, Os, Ir, Pt).
- 03.C Aluminofluorides, 06 Borates, 08 Vanadates (04.H $V^{[5,6]}$ Vanadates), 09 Silicates:
    - Neso: insular.
    - Soro: grouping.
    - Cyclo: ring.
    - Ino: chain.
    - Phyllo: sheet.
    - Tecto: three-dimensional framework.

- Nickel–Strunz code scheme: NN.XY.##x:
    - NN: Nickel–Strunz mineral class number.
    - X: Nickel–Strunz mineral division letter.
    - Y: Nickel–Strunz mineral family letter.
    - ##x: Nickel–Strunz mineral/group number, x add-on letter.

Class: Native Elements

- 01.A Metals and Intermetallic Alloys:
    - 01.AA Copper-cupalite family: 05 Native copper, 05 Lead, 05 Native gold, 05 Native silver, 05 Nickel, 05 Aluminium; 10a Auricupride, 10b Tetra-auricupride; 15 Novodneprite, 15 Khatyrkite, 15 Anyuiite; 20 Cupalite, 25 Hunchunite.
    - 01.AB Zinc-brass family (Cu-Zn alloys): 05 Cadmium, 05 Zinc, 05 Titanium*, 05 Rhenium*; 10a Brass*, 10a Zhanghengite, 10b Danbaite, 10b Tongxinite*.
    - 01.AC Indium-tin family: 05 Indium, 10 Tin; 15 Yuanjiangite, 15 Sorosite.
    - 01.AD Mercury-amalgam family: 00 Amalgam*, 05 Mercury; 10 Belendorffite, 10 Kolymite; 15a Paraschachnerite, 15a Schachnerite, 15b Luanheite, 15c Eugenite, 15d Moschellandsbergite; 20a Weishanite, 20b Goldamalgam*; 25 Potarite, 30 Leadamalgam.
    - 01.AE Iron-chromium family: 05 Kamacite? (Iron var.), 05 Iron, 05 Chromium; 10 Antitaenite*, 10 Taenite, 10 Tetrataenite; 15 Chromferide, 15 Wairauite, 15 Ferchromide; 20 Awaruite, 25 Jedwabite.
    - 01.AF Platinum-group elements: 05 Osmium, 05 Rutheniridosmine, 05 Ruthenium; 10 Palladium, 10 Iridium, 10 Rhodium, 10 Platinum.
    - 01.AG PGE-metal alloys: 05 Garutiite, 05 Hexaferrum; 10 Atokite, 10 Zvyagintsevite, 10 Rustenburgite; 15 Taimyrite, 15 Tatyanaite; 20 Paolovite; 25 Plumbopalladinite, 25 Stannopalladinite; 30 Cabriite; 35 Chengdeite, 35 Isoferroplatinum; 40 Ferronickelplatinum, 40 Tetraferroplatinum, 40 Tulameenite; 45 Hongshiite*, 45 Skaergaardite; 50 Yixunite, 55 Damiaoite, 60 Niggliite, 65 Bortnikovite, 70 Nielsenite.

- 01.B Metallic Carbides, Silicides, Nitrides and Phosphides:
  - 01.BA Carbides: 05 Cohenite; 10 Isovite, 10 Haxonite; 15 Tongbaite; 20 Khamrabaevite, 20 Niobocarbide, 20 Tantalcarbide; 25 Qusongite, 30 Yarlongite.
  - 01.BB Silicides: Zangboite; 05 Mavlyanovite, 05 Suessite; 10 Perryite, 15 Fersilicite*, 20 Ferdisilicite*, 25 Luobusaite, 30 Gupeiite, 35 Hapkeite, 40 Xifengite.
  - 01.BC Nitrides: 05 Roaldite, 10 Siderazot, 15 Carlsbergite, 15 Osbornite.
  - 01.BD Phosphides: 05 Schreibersite, 05 Nickelphosphide; 10 Barringerite, 10 Monipite; 15 Allabogdanite, 15 Florenskyite, 15 Andreyivanovite; 20 Melliniite.
- 01.C Metalloids and Nonmetals:
  - 01.CA Arsenic group elements: 05 Bismuth, 05 Antimony, 05 Arsenic, 05 Stibarsen; 10 Arsenolamprite, 10 Pararsenolamprite; 15 Paradocrasite.
  - 01.CB Carbon-silicon family: 05a Graphite, 05b Chaoite, 05c Fullerite; 10a Diamond, 10b Lonsdaleite, 15 Silicon.
  - 01.CC Sulfur-selenium-iodine: 05 Sulfur, 05 Rosickyite; 10 Tellurium, 10 Selenium.
- 01.D Nonmetallic Carbides and Nitrides:
  - 01.DA Nonmetallic carbides: 05 Moissanite.
  - 01.DB Nonmetallic nitrides: 05 Nierite, 10 Sinoite.
- 01.X Unclassified Strunz Elements (Metals and intermetallic alloys; metalloids and nonmetals; carbides, silicides, nitrides, phosphides):
  - 01.XX Unknown: 00 Hexamolybdenum, 00 Tantalum*, 00 Brownleeite.

Oxide Minerals

The oxide mineral class includes those minerals in which the oxide anion (O^{2-}) is bonded to one or more metal ions. The hydroxide bearing minerals are typically included in the oxide class. The minerals with complex anion groups such as the silicates, sulfates, carbonates and phosphates are classed separately.

Oxide mineral exhibit at the Museum of Geology in South Dakota.

Simple oxides:

- X_2O and XO:
    - Cuprite Cu_2O.
    - Ice H_2O.
    - Periclase group:
        - Periclase MgO.
        - Manganosite MnO.
    - Zincite group:
        - Zincite ZnO.
        - Bromellite BeO.
        - Tenorite CuO.
        - Litharge PbO.
- X_2O_3:
    - Hematite group:
        - Corundum Al_2O_3.
        - Hematite Fe_2O_3.
        - Ilmenite $FeTiO_3$.
- XO_2:
    - Rutile group:
        - Rutile TiO_2.
        - Pyrolusite MnO_2.
        - Cassiterite SnO_2.
    - Baddeleyite ZrO_2.
    - Uraninite UO_2.
    - Thorianite ThO_2.
- XY_2O_4:
    - Spinel group:
        - Spinel $MgAl_2O_4$.
        - Gahnite $ZnAl_2O_4$.
        - Magnetite Fe_3O_4 ($Fe^{2+}Fe^{3+}{}_2O_4$).
        - Franklinite $(Zn,Fe,Mn)(Fe,Mn)_2O_4$.

        - Chromite $FeCr_2O_4$.
    - Chrysoberyl $BeAl_2O_4$.
    - Columbite$(Fe,Mn)(Nb,Ta)_2O_6$.

Hydroxide subgroup:

- Brucite $Mg(OH)_2$.
- Manganite MnO(OH).
- Romanechite $BaMn^{2+}Mn^{4+}{}_8O_{16}(OH)_4$.
- Goethite group:
    - Diaspore αAlO(OH).
    - Goethite αFeO(OH).

Nickel–Strunz Classification -04- Oxides

IMA-CNMNC proposes a new hierarchical scheme (Mills et al., 2009). This list uses it to modify the Classification of Nickel–Strunz (mindat.org, 10 ed, pending publication).

- Abbreviations:
    - "*" - discredited (IMA/CNMNC status).
    - "?" - questionable/doubtful (IMA/CNMNC status).
    - "REE" - Rare-earth element (Sc, Y, La, Ce, Pr, Nd, Pm, Sm, Eu, Gd, Tb, Dy, Ho, Er, Tm, Yb, Lu).
    - "PGE" - Platinum-group element (Ru, Rh, Pd, Os, Ir, Pt).
    - 03.C Aluminofluorides, 06 Borates, 08 Vanadates (04.H $V^{[5,6]}$ Vanadates), 09 Silicates:
        - Neso: insular.
        - Soro: grouping.
        - Cyclo: ring.
        - Ino: chain.
        - Phyllo: sheet.
        - Tekto: three-dimensional framework.
- Nickel–Strunz code scheme: NN.XY.##x.
    - NN: Nickel–Strunz mineral class number.
    - X: Nickel–Strunz mineral division letter.
    - Y: Nickel–Strunz mineral family letter.

- ##x: Nickel–Strunz mineral/group number, x add-on letter.

Class: Oxides

- 04.A Metal:Oxygen = 2.1 and 1:1:
  - 04.AA Cation:Anion (M:O) = 2:1 (and 1.8:1): 05 Ice, 10 Cuprite, 15 Paramelaconite.
  - 04.AB M:O = 1:1 (and up to 1:1.25); with small to medium-sized cations only: 05 Crednerite, 10 Tenorite; 15 Delafossite, 15 Mcconnellite; 20 Bromellite, 20 Zincite; 25 Lime, 25 Bunsenite, 25 Monteponite, 25 Manganosite, 25 Periclase, 25 Wustite.
  - 04.AC M:O = 1:1 (and up to 1:1.25); with large cations (± smaller ones): 05 Swedenborgite; 10 Brownmillerite, 10 Srebrodolskite; 15 Montroydite, 20 Litharge, 20 Romarchite, 25 Massicot.
- 04.B Metal:Oxygen = 3:4 and similar:
  - 04.BA With small and medium-sized cations: 05 Chrysoberyl, 10 Manganostibite.
  - 04.BB With only medium-sized cations: 05 Filipstadite, 05 Donathite?, 05 Gahnite, 05 Galaxite, 05 Hercynite, 05 Spinel, 05 Cochromite, 05 Chromite, 05 Magnesiochromite, 05 Manganochromite, 05 Nichromite, 05 Zincochromite, 05 Magnetite, 05 Cuprospinel, 05 Franklinite, 05 Jacobsite, 05 Magnesioferrite, 05 Trevorite, 05 Brunogeierite, 05 Coulsonite, 05 Magnesiocoulsonite, 05 Qandilite, 05 Ulvospinel, 05 Vuorelainenite; 10 Hydrohetaerolite, 10 Hausmannite, 10 Iwakiite, 10 Hetaerolite; 15 Maghemite, 20 Tegengrenite, 25 Xieite.
  - 04.BC With medium-sized and large cations: 05 Marokite, 10 Dmitryivanovite.
  - 04.BD With only large cations: 05 Minium.

04.C Metal:Oxygen = 2:3, 3:5, and Similar:

  - 04.CB With medium-sized cations: 05 Tistarite, 05 Auroantimonate*, 05 Brizziite-VII, 05 Brizziite-III, 05 Corundum, 05 Eskolaite, 05 Hematite, 05 Karelianite, 05 Geikielite, 05 Ecandrewsite, 05 Ilmenite, 05 Pyrophanite, 05 Melanostibite, 05 Romanite*; 10 Bixbyite, 10 Avicennite; 15 Armalcolite, 15 Mongshanite*, 15 Pseudobrookite; 20 Magnesiohogbomite-6N6S, 20 Magnesiohogbomite-2N3S, 20 Magnesiohogbomite-2N2S, 20 Zincohogbomite-2N2S, 20 Ferrohogbomite-2N2S; 25 Pseudorutile, 25 Ilmenorutile; 30 Oxyvanite, 30 Berdesinskiite; 35 Olkhonskite, 35 Schreyerite; 40 Kamiokite, 40 Nolanite, 40 Rinmanite; 45 Stibioclaudetite, 45 Claudetite; 50 Arsenolite, 50 Senarmontite; 55 Valentinite, 60 Bismite, 65 Sphaerobismoite, 70 Sillenite, 75 Kyzylkumite.
  - 04.CC With large and medium-sized cations: 05 Chrombismite, 10 Freudenbergite, 15 Grossite, 20 Mayenite, 25 Yafsoanite; 30 Barioperovskite, 30 Lakargiite, 30 Natroniobite, 30 Latrappite, 30 Lueshite, 30 Perovskite; 35 Macedonite, 35 Isolueshite, 35 Loparite-(Ce), 35 Tausonite; 40 Crichtonite, 40 Dessauite, 40 Davidite-(Ce), 40 Davidite-(La), 40 Mathiasite, 40 Lindsleyite, 40 Landauite, 40 Loveringite, 40 Loveringite, 40 Cleusonite, 40 Gramaccioliite-(Y); 45 Hawthorneite, 45 Magnetoplumbite, 45 Haggertyite, 45 Batiferrite, 45 Hibonite, 45 Nezilovite, 45 Yimengite, 45 Diaoyudaoite,

45 Lindqvistite, 45 Plumboferrite; 50 Jeppeite, 55 Zenzenite, 60 Mengxianminite*.

04.D Metal:Oxygen = 1:2 and similar:

- 04.DA With small cations.
- (moved to -09- Subclass: tektosilicates).
- 04.DB With medium-sized cations; chains of edge-sharing octahedra: 05 Tripuhyite, 05 Tugarinovite, 05 Varlamoffite*, 05 Argutite, 05 Cassiterite, 05 Rutile, 05 Pyrolusite, 05 Plattnerite, 05 Squawcreekite?; 10 Bystromite, 10 Ordonezite, 10 Tapiolite-(Fe), 10 Tapiolite-(Mn), 10 Tapiolite*, 15a Paramontroseite, 15a Ramsdellite, 15b Akhtenskite, 15c Nsutite; 20 Scrutinyite; 25 Ixiolite, 25 Ishikawaite, 25 Srilankite, 25 Samarskite-(Y), 25 Samarskite-(Yb), 25 Yttrocolumbite-(Y); 30 Heftetjernite, 30 Wolframoixiolite*, 30 Krasnoselskite*, 30 Ferberite, 30 Hubnerite, 30 Sanmartinite, 30 Wolframite*; 35 Tantalite-(Mg), 35 Tantalite-(Fe), 35 Tantalite-(Mn), 35 Columbite-(Mg), 35 Columbite-(Fe), 35 Columbite-(Mn), 35 Qitianlingite; 40 Ferrowodginite, 40 Lithiotantite, 40 Lithiowodginite, 40 Tantalowodginite*, 40 Titanowodginite, 40 Wodginite, 40 Ferrotitanowodginite; 45 Tivanite, 50 Carmichaelite, 55 Alumotantite, 60 Biehlite.
- 04.DC With medium-sized cations; sheets of edge-sharing octahedra: 05 Bahianite, 10 Simpsonite.
- 04.DD With medium-sized cations; frameworks of edge-sharing octahedra: 05 Anatase, 10 Brookite.
- 04.DE With medium-sized cations; with various polyhedra: 05 Downeyite, 10 Koragoite; 15 Koechlinite, 15 Russellite, 15 Tungstibite; 20 Tellurite, 25 Paratellurite; 30 Cervantite, 30 Bismutotantalite, 30 Bismutocolumbite, 30 Clinocervantite, 30 Stibiocolumbite, 30 Stibiotantalite; 35 IMA2007-058, 35 Baddeleyite.
- 04.DF With large (± medium-sized) cations; dimers and trimers of edge-sharing octahedra: 05 Nioboaeschynite-(Y), 05 Aeschynite-(Ce), 05 Aeschynite-(Nd), 05 Aeschynite-(Y), 05 Nioboaeschynite-(Ce), 05 Nioboaeschynite-(Nd), 05 Tantalaeschynite-(Y), 05 Rynersonite, 05 Vigezzite, 10 Changbaiite, 15 Murataite.
- 04.DG With large (± medium-sized) cations; chains of edge-sharing octahedra: 05 Euxenite-(Y), 05 Loranskite-(Y), 05 Polycrase-(Y), 05 Uranopolycrase, 05 Fersmite, 05 Kobeite-(Y), 05 Tanteuxenite-(Y), 05 Yttrocrasite-(Y); 10 Fergusonite-beta-(Nd), 10 Fergusonite-beta-(Y), 10 Fergusonite-beta-(Ce), 10 Yttrotantalite-(Y); 15 Foordite, 15 Thoreaulite; 20 Raspite.
- 04.DH With large (± medium-sized) cations; sheets of edge-sharing octahedra:
  - IMA/CNMNC revised the Pyrochlore supergroup 2010 (04.DH.15 and 04.DH.20).
  - 05 Brannerite, 05 Orthobrannerite, 05 Thorutite; 10 Kassite, 10 Lucasite-(Ce).
  - Pyrochlore group: Fluorcalciopyrochlore, Fluorkenopyrochlore, Fluornatropyrochlore, Fluorstrontiopyrochlore, Hydropyrochlore, Hydroxycalciopyrochlore,

Kenoplumbopyrochlore, Oxycalciopyrochlore, Oxynatropyrochlore, Oxyplumbopyrochlore, Oxyyttropyrochlore-(Y).

- Microlite group: Fluorcalciomicrolite, Fluornatromicrolite, Hydrokenomicrolite, Hydromicrolite, Hydroxykenomicrolite, Kenoplumbomicrolite, Oxycalciomicrolite, Oxystannomicrolite, Oxystibiomicrolite.
- Romeite group: Cuproromeite, Fluorcalcioromeite, Fluornatroromeite, Hydroxycalcioromeite, Oxycalcioromeite, Oxyplumboromeite, Stibiconite.
- Betafite group: Calciobetafite, Oxyuranobetafite.
- Elsmoreite group: Hydrokenoelsmoreite.
- 25 Rosiaite; 30 Zirconolite-3O, 30 Zirconolite-3T, 30 Zirconolite-2M, 30 Zirconolite; 35 Liandratite, 35 Petscheckite; 40 Ingersonite, 45 Pittongite.
- Discredited minerals 04.DH.15: Bariomicrolite (of Hogarth 1977), Bariopyrochlore (of Hogarth 1977), Betafite (of Hogarth 1977), Bismutomicrolite (of Hogarth 1977), Ceriopyrochlore (of Hogarth 1977), Jixianite, Natrobistantite, Plumbomicrolite (of Hogarth 1977), Plumbobetafite (of Hogarth 1977), Stannomicrolite (of Hogarth 1977), Stibiobetafite (of Černý et al.), Yttrobetafite (of Hogarth 1977), Yttropyrochlore (of Hogarth 1977), Bismutopyrochlore (of Chukanov et al.) and Bismutostibiconite 04.DH.20.

- 04.DJ With large (± medium-sized) cations; polyhedral frameworks: 05 Calciotantite, 05 Irtyshite, 05 Natrotantite.
- 04.DK With large (± medium-sized) cations; tunnel structures: 05 Ankangite, 05 Coronadite, 05 Hollandite, 05 Manjiroite, 05 Mannardite, 05 Redledgeite, 05 Priderite, 05 Henrymeyerite, 05 Akaganeite, 10 Cryptomelane, 10 Romanechite, 10 Strontiomelane, 10 Todorokite.
- 04.DL With large (± medium-sized) cations; fluorite-type structures: 05 Cerianite-(Ce), 05 Zirkelite, 05 Thorianite, 05 Uraninite; 10 Calzirtite, 10 Hiarneite, 10 Tazheranite.
- 04.DM With large (± medium-sized) cations; unclassified: 05 Sosedkoite, 05 Rankamaite; 15 Cesplumtantite, 20 Eyselite, 25 Kuranakhite.

- 04.E Metal:Oxygen = < 1:2:
  - 04.E: IMA2008-040.
  - 04.EA Oxides with metal : oxygen < 1:2 (M_2O_5, MO_3): 05 Tantite, 10 Krasnogorite*, 10 Molybdite.
- 04.X Unclassified Strunz Oxides:
  - 04.XX Unknown: 00 Allendeite, 00 Ashanite?, 00 Hongquiite*, 00 Psilomelane?, 00 Uhligite?, 00 Clinobirnessite*, 00 Kleberite*, 00 Chubutite*, 00 Struverite?, 00 IMA2000-016, 00 IMA2000-026.

Class: Hydroxides

- 04.F Hydroxides (without V or U):
    - 04.FA Hydroxides with OH, without H_2O; corner-sharing tetrahedra: 05a Behoite, 05b Clinobehoite; 10 Sweetite, 10 Wulfingite, 10 Ashoverite.
    - 04.FB Hydroxides with OH, without H_2O; insular octahedra: 05 Shakhovite; 10 Cualstibite, 10 Zincalstibite.
    - 04.FC Hydroxides with OH, without H_2O; corner-sharing octahedra: 05 Dzhalindite, 05 Sohngeite, 05 Bernalite; 10 Burtite, 10 Mushistonite, 10 Natanite, 10 Vismirnovite, 10 Schoenfliesite, 10 Wickmanite; 15 Jeanbandyite, 15 Mopungite, 15 Stottite; 15 Tetrawickmanite; 20 Ferronigerite-6N6S, 20 Ferronigerite-2N1S, 20 Magnesionigerite-6N6S, 20 Magnesionigerite-2N1S; 25 Magnesiotaaffeite-6N3S, 25 Magnesiotaaffeite-2N2S, 25 Ferrotaaffeite-6N3S.
    - 04.FD Hydroxides with OH, without H_2O; chains of edge-sharing octahedra: 05 Spertiniite; 10 Bracewellite, 10 Diaspore, 10 Guyanaite, 10 Groutite, 10 Goethite, 10 Montroseite, 10 Tsumgallite; 15 Manganitev; 20 Cerotungstite-(Ce), 20 Yttrotungstite-(Y), 20 Yttrotungstite-(Ce); 25 Frankhawthorneite; 30 Khinite, 30 Parakhinite.
    - 04.FE Hydroxides with OH, without H_2O; sheets of edge-sharing octahedra: 05 Amakinite, 05 Brucite, 05 Portlandite, 05 Pyrochroite, 05 Theophrastite, 05 Fougerite; 10 Bayerite, 10 Doyleite, 10 Gibbsite, 10 Nordstrandite; 15 Boehmite, 15 Lepidocrocite; 20 Grimaldiite, 20 Heterogenite-2H, 20 Heterogenite-3R; 25 Feitknechtite, 25 Lithiophorite; 30 Quenselite, 35 Ferrihydrite; 40 Feroxyhyte, 40 Vernadite; 45 Quetzalcoatlite.
    - 04.FF Hydroxides with OH, without H_2O; various polyhedra: 05 Hydroromarchite.
    - 04.FG Hydroxides with OH, without H_2O; unclassified: 05 Janggunite, 10 Cesarolite, 15 Kimrobinsonite.
    - 04.FH Hydroxides with H_2O ± (OH); insular octahedra: 05 Bottinoite, 05 Brandholzite.
    - 04.FJ Hydroxides with H_2O ± (OH); corner-sharing octahedra: 05 Sidwillite, 05 Meymacite; 10 Tungstite; 15 Ilsemannite, 15 Hydrotungstite; 20 Parabariomicrolite.
    - 04.FK Hydroxides with H_2O ± (OH); chains of edge-sharing octahedra: 05 Bamfordite.
    - 04.FL Hydroxides with H_2O ± (OH); sheets of edge-sharing octahedra: 05 Meixnerite, 05 Jamborite, 05 Iowaite, 05 Woodallite, 05 Akdalaite, 05 Muskoxite; 10 Hydrocalumite, 15 Kuzelite; 20 Aurorite, 20 Chalcophanite, 20 Ernienickelite, 20 Jianshuiite; 25 Woodruffite, 30 Asbolane; 40 Takanelite, 40 Rancieite; 45 Birnessite, 55 Cianciulliite, 60 Jensenite, 65 Leisingite, 75 Cafetite, 80 Mourite, 85 Deloryite.
    - 04.FM Hydroxides with H_2O ± (OH); Unclassified: 15 Franconite, 15 Hochelagaite, 15 Ternovite; 25 Belyankinite, 25 Gerasimovskite, 25 Manganbelyankinite; 30 Silhydrite, 35 Cuzticite, 40 Cyanophyllite.
    - 04.FN: 05 Menezesite.

- 04.G Uranyl Hydroxides:
  - 04.GA Without additional cations: IMA2008-022; 05 Metaschoepite, 05 Paraschoepite, 05 Schoepite; 10 Ianthinite; 15 Metastudtite, 15 Studtite.
  - 04.GB With additional cations (K, Ca, Ba, Pb, etc.); with mainly $UO_2(O,OH)_5$ pentagonal polyhedra: 05 Compreignacite, 05 Agrinierite, 05 Rameauite; 10 Billietite, 10 Becquerelite, 10 Protasite; 15 Richetite; 20 Calciouranoite, 20 Bauranoite, 20 Metacalciouranoite; 25 Fourmarierite, 30 Wolsendorfite, 35 Masuyite; 40 Metavandendriesscheite, 40 Vandendriesscheite; 45 Vandenbrandeite, 50 Sayrite, 55 Curite, 60 Iriginite, 65 Uranosphaerite, 70 Holfertite.
  - 04.GC With additional cations; with $UO_2(O,OH)_6$ hexagonal polyhedra: 05 Clarkeite, 10 Umohoite, 15 Spriggite.
- 04.H V Vanadates:
  - (moved to -08- Class: vanadates).
- 04.I Ice group.
- 04.X Unclassified Strunz Oxides (Hydroxides):
  - 04.XX Unknown: 00 Ungursaite*, 00 Scheteligite?

Halide Minerals

The halide mineral class include those minerals with a dominant halide anion (F^-, Cl^-, Br^- and I^-). Complex halide minerals may also have polyatomic anions in addition to or that include halides.

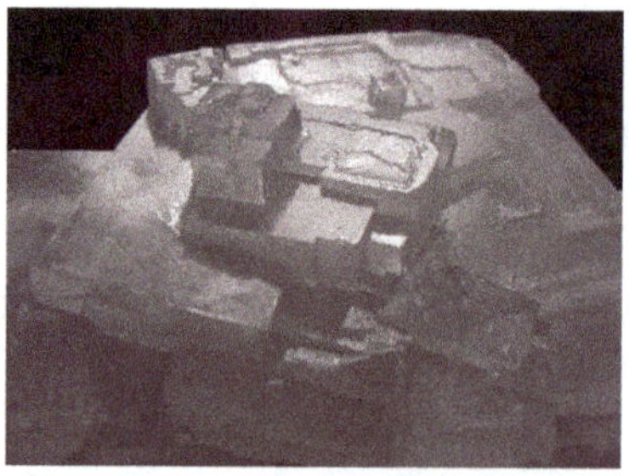

Halite

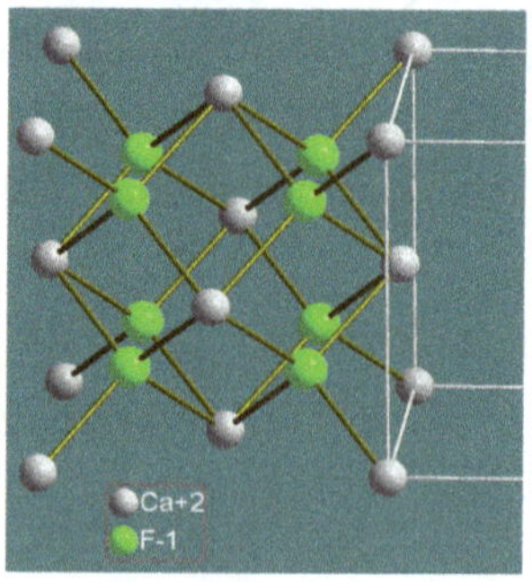

Fluorite structure

Examples include the following:

- Halite NaCl.
- Sylvite KCl.
- Chlorargyrite AgCl and bromargyrite AgBr.
- Fluorite CaF_2.
- Atacamite $Cu_2Cl(OH)_3$.
- Bischofite ($MgCl_2 \cdot 6H_2O$).
- Carnallite $KMgCl_3 \cdot 6H_2O$.
- Cryolite Na_3AlF_6.
- Cryptohalite (α)$(NH_4)_2SiF_6$.
- Bararite (β)$(NH_4)_2SiF_6$.

Commercially Significant Halide Minerals

Two commercially important halide minerals are halite and fluorite. The former is a major source of sodium chloride, in parallel with sodium chloride extracted from sea water or brine wells. Fluorite is a major source of hydrogen fluoride, complementing the supply obtained as a byproduct of the production of fertilizer. Carnallite and bischofite are important sources of magnesium. Natural cryolite was historically required for the production of aluminium, however, currently most cryolite used is produced synthetically.

Many of the halide minerals occur in marine evaporite deposits. The Atacama Desert also has large quantities of halide minerals as well as chlorates, iodates, oxyhalides and the like as well as nitrates, borates and other water-soluble minerals—not only underground but it crusts on the surface due to the low rainfall—the Atacama is the world's driest desert as well as one of the oldest (25 million years).

Nickel–Strunz Classification -03- Halides

IMA-CNMNC proposes a new hierarchical scheme (Mills et al., 2009). This list uses the Classification of Nickel–Strunz (mindat.org, 10 ed, pending publication).

- Abbreviations:
  - "*" - discredited (IMA/CNMNC status).
  - "?" - questionable/doubtful (IMA/CNMNC status).
  - "REE" - Rare-earth element (Sc, Y, La, Ce, Pr, Nd, Pm, Sm, Eu, Gd, Tb, Dy, Ho, Er, Tm, Yb, Lu).
  - "PGE" - Platinum-group element (Ru, Rh, Pd, Os, Ir, Pt).
  - 03.C Aluminofluorides, 06 Borates, 08 Vanadates (04.H $V^{[5,6]}$ Vanadates), 09 Silicates:
    - Neso: insular.

- Soro: grouping.
- Cyclo: ring.
- Ino: chain.
- Phyllo: sheet.
- Tekto: three-dimensional framework.

- Nickel–Strunz code scheme: NN.XY.##x:
    - NN: Nickel–Strunz mineral class number.
    - X: Nickel–Strunz mineral division letter.
    - Y: Nickel–Strunz mineral family letter.
    - ##x: Nickel–Strunz mineral/group number, x add-on letter.

Class: Halides

Halide specimens at Museum of Geology, South Dakota.

- 03.A Simple halides, without H_2O.
    - 03.AA M:X = 1:1, 2:3, 3:5, etc.: Panichiite; 05 Nantokite, 05 Marshite, 05 Miersite; 10 Iodargyrite, 10 Tocornalite; 15 Bromargyrite, 15 Embolite*, 15 Chlorargyrite; 20 Carobbiite, 20 Griceite, 20 Halite, 20 Sylvite, 20 Villiaumite; 25 Salammoniac, 25 Lafossaite; 30 Calomel, 30 Kuzminite, 30 Moschelite; 35 Neighborite; 40 Chlorocalcite, 45 Kolarite, 50 Radhakrishnaite; 55 Hephaistosite, 55 Challacolloite.
    - 03.AB M:X = 1:2: 05 Tolbachite, 10 Coccinite, 15 Sellaite; 20 Chloromagnesite*, 20 Lawrencite, 20 Scacchite; 25 Frankdicksonite, 25 Fluorite; 30 Tveitite-(Y); 35 Gagarinite-(Y); 35 Zajacite-(Ce).
    - 03.AC M:X = 1:3: 05 Zharchikhite, 10 Molysite; 15 Fluocerite-(Ce), 15 Fluocerite-(La), 20 Gananite.
- 03.B Simple Halides, with H_2O:
    - 03.BA M:X = 1:1 and 2:3: 05 Hydrohalite, 10 Carnallite.

- 03.BB M:X = 1:2: 05 Eriochalcite, 10 Rokuhnite, 15 Bischofite, 20 Nickelbischofite, 25 Sinjarite, 30 Antarcticite, 35 Tachyhydrite.
- 03.BC M:X = 1:3: 05 Chloraluminite.
- 03.BD Simple Halides with H_2O and additional OH: 05 Cadwaladerite, 10 Lesukite, 15 Korshunovskite, 20 Nepskoeite, 25 Koenenite.

- 03.C Complex Halides:
  - 03.C: Steropesite, IMA2008-032, IMA2008-039.
  - 03.CA Borofluorides: 05 Ferruccite; 10 Avogadrite, 10 Barberiite.
  - 03.CB Neso-aluminofluorides: 05 Cryolithionite; 15 Cryolite, 15 Elpasolite, 15 Simmonsite; 20 Colquiriite, 25 Weberite, 30 Karasugite, 35 Usovite; 40 Pachnolite, 40 Thomsenolite; 45 Carlhintzeite, 50 Yaroslavite.
  - 03.CC Soro-aluminofluorides: 05 Gearksutite; 10 Acuminite, 10 Tikhonenkovite; 15 Artroeite; 20 Calcjarlite, 20 Jarlite, 20 Jorgensenite.
  - 03.CD Ino-aluminofluorides: 05 Rosenbergite, 10 Prosopite.
  - 03.CE Phyllo-aluminofluorides: 05 Chiolite.
  - 03.CF Tekto-aluminofluorides: 05 Ralstonite, 10 Boldyrevite?, 15 Bogvadite.
  - 03.CG Aluminofluorides with CO_3, SO_4, PO_4: 05 Stenonite; 10 Chukhrovite-(Nd), 10 Chukhrovite-(Ce), 10 Chukhrovite-(Y), 10 Meniaylovite; 15 Creedite, 20 Boggildite, 25 Thermessaite.
  - 03.CH: 05 Malladrite, 10 Bararite; 15 Cryptohalite, 15 Hieratite; 20 Demartinite, 25 Knasibfite.
  - 03.CJ With MX_6 complexes; M = Fe, Mn, Cu: 05 Chlormanganokalite, 05 Rinneite; 10 Erythrosiderite, 10 Kremersite; 15 Mitscherlichite, 20 Douglasite, 30 Zirklerite.
- 03.D Oxyhalides, Hydroxyhalides and Related Double Halides:
  - 03.DA With Cu, etc., without Pb: 05 Melanothallite; 10a Atacamite, 10a Kempite, 10a Hibbingite, 10b Botallackite, 10b Clinoatacamite, 10b Belloite, 10c Gillardite, 10c Kapellasite, 10c Haydeeite, 10c Paratacamite, 10c Herbertsmithite; 15 Claringbullite, 20 Simonkolleite; 25 Buttgenbachite, 25 Connellite; 30 Abhurite, 35 Ponomarevite; 40 Calumetite, 40 Anthonyite; 45 Khaidarkanite, 50 Bobkingite, 55 Avdoninite, 60 Droninoite.
  - 03.DB With Pb, Cu, etc.: 05 Diaboleite, 10 Pseudoboleite, 15 Boleite, 20 Cumengite, 25 Bideauxite, 30 Chloroxiphite, 35 Hematophanite; 40 Asisite, 40 Parkinsonite; 45 Murdochite, 50 Yedlinite.
  - 03.DC With Pb (As, Sb, Bi), without Cu: 05 Laurionite, 05 Paralaurionite; 10 Fiedlerite, 15 Penfieldite, 20 Laurelite; 25 Zhangpeishanite, 25 Matlockite, 25 Rorisite, 25 Daubreeite, 25 Bismoclite, 25 Zavaritskite; 30 Nadorite, 30 Perite; 35 Aravaipaite, 37

Calcioaravaipaite, 40 Thorikosite, 45 Mereheadite, 50 Blixite, 55 Pinalite, 60 Symesite; 65 Ecdemite, 65 Heliophyllite; 70 Mendipite, 75 Damaraite, 80 Onoratoite, 85 Cotunnite, 90 Pseudocotunnite, 95 Barstowite.

- 03.DD With Hg: 05 Eglestonite, 05 Kadyrelite; 10 Poyarkovite, 15 Hanawaltite, 20 Terlinguaite, 25 Pinchite; 30 Mosesite, 30 Gianellaite; 35 Kleinite, 40 Tedhadleyite, 45 Vasilyevite, 50 Aurivilliusite, 55 Terlinguacreekite, 60 Kelyanite, 65 Comancheite.
- 03.DE With Rare-Earth Elements: 05 Haleniusite-(La).

- 03.X Unclassified Strunz Halogenides:
  - 03.XX Unknown: 00 Hydrophilite?, 00 Hydromolysite?, 00 Yttrocerite*, 00 Lorettoite?, 00 IMA2009-014, 00 IMA2009-015.

Sulfate Minerals

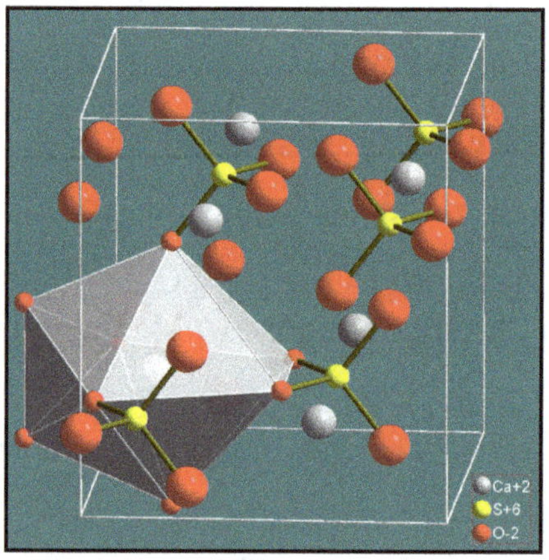

Anhydrite crystal structure.

The sulfate minerals are a class of minerals which include the sulfate ion (SO_4^{2-}) within their structure. The sulfate minerals occur commonly in primary evaporite depositional environments, as gangue minerals in hydrothermal veins and as secondary minerals in the oxidizing zone of sulfide mineral deposits. The chromate and manganate minerals have a similar structure and are often included with the sulfates in mineral classification systems.

Barite with cerussite.

Sulfate minerals include:

- Anhydrous sulfates:
    - Barite $BaSO_4$.
    - Celestite $SrSO_4$.
    - Anglesite $PbSO_4$.
    - Anhydrite $CaSO_4$.
    - Hanksite $Na_{22}K(SO_4)_9(CO_3)_2Cl$.
- Hydroxide and hydrous sulfates:
    - Gypsum $CaSO_4{\cdot}2H_2O$.
    - Chalcanthite $CuSO_4{\cdot}5H_2O$.
    - Kieserite $MgSO_4{\cdot}H_2O$.
    - Starkeyite $MgSO_4{\cdot}4H_2O$.
    - Hexahydrite $MgSO_4{\cdot}6H_2O$.
    - Epsomite $MgSO_4{\cdot}7H_2O$.
    - Meridianiite $MgSO_4{\cdot}11H_2O$.
    - Melanterite $FeSO_4{\cdot}7H_2O$.
    - Antlerite $Cu_3SO_4(OH)_4$.
    - Brochantite $Cu_4SO_4(OH)_6$.
    - Alunite $KAl_3(SO_4)_2(OH)_6$.
    - Jarosite $KFe_3(SO_4)_2(OH)_6$.

Nickel–Strunz Classification -07- Sulfates

Hanksite, one of the rare minerals that is a sulfate and carbonate.

IMA-CNMNC proposes a new hierarchical scheme (Mills et al., 2009). This list uses it to modify the Classification of Nickel–Strunz (mindat.org, 10 ed, pending publication).

- Abbreviations:
  - "*" - discredited (IMA/CNMNC status).
  - "?" - questionable/doubtful (IMA/CNMNC status).
  - "REE" - Rare-earth element (Sc, Y, La, Ce, Pr, Nd, Pm, Sm, Eu, Gd, Tb, Dy, Ho, Er, Tm, Yb, Lu).
  - "PGE" - Platinum-group element (Ru, Rh, Pd, Os, Ir, Pt).
  - 03.C Aluminofluorides, 06 Borates, 08 Vanadates (04.H $V^{[5,6]}$ Vanadates), 09 Silicates:
    - Neso: insular.
    - Soro: grouping.
    - Cyclo: ring.
    - Ino: chain.
    - Phyllo: sheet.
    - Tekto: three-dimensional framework.
- Nickel–Strunz code scheme: NN.XY.##x:
  - NN: Nickel–Strunz mineral class number.
  - X: Nickel–Strunz mineral division letter.
  - Y: Nickel–Strunz mineral family letter.
  - ##x: Nickel–Strunz mineral/group number, x add-on letter.

Class: Sulfates, Selenates, Tellurates

- 07.A Sulfates (selenates, etc.) without Additional Anions, without H_2O:
  - 07.AB With medium-sized cations: 05 Millosevichite, 05 Mikasaite; 10 Chalcocyanite, 10 Zincosite*.
  - 07.AC With medium-sized and large cations: IMA2008-029, 05 Vanthoffite; 10 Efremovite, 10 Manganolangbeinite, 10 Langbeinite; 15 Eldfellite, 15 Yavapaiite; 20 Godovikovite, 20 Sabieite; 25 Thenardite, 35 Aphthitalite.
  - 07.AD With only large cations: 05 Arcanite, 05 Mascagnite; 10 Mercallite, 15 Misenite, 20 Letovicite, 25 Glauberite, 30 Anhydrite; 35 Anglesite, 35 Barite, 35 Celestine, 35 Radiobarite*, 35 Olsacherite; 40 Kalistrontite, 40 Palmierite.
- 07.B Sulfates (selenates, etc.) with Additional Anions, without H_2O:
  - 07.BB With medium-sized cations: 05 Caminite, 10 Hauckite, 15 Antlerite, 20 Dolerophanite, 25 Brochantite, 30 Vergasovaite, 35 Klebelsbergite, 40 Schuetteite, 45 Paraotwayite, 50 Xocomecatlite, 55 Pauflerite.

  - 07.BC With medium-sized and large cations: 05 Dansite; 10 Alunite, 10 Ammonioalunite, 10 Ammoniojarosite, 10 Beaverite, 10 Argentojarosite, 10 Huangite, 10 Dorallcharite, 10 Jarosite, 10 Hydroniumjarosite, 10 Minamiite, 10 Natrojarosite, 10 Natroalunite, 10 Osarizawaite, 10 Plumbojarosite, 10 Walthierite, 10 Schlossmacherite; 15 Yeelimite; 20 Atlasovite, 20 Nabokoite; 25 Chlorothionite; 30 Fedotovite, 30 Euchlorine; 35 Kamchatkite, 40 Piypite; 45 Klyuchevskite-Duplicate, 45 Klyuchevskite, 45 Alumoklyuchevskite; 50 Caledonite, 55 Wherryite, 60 Mammothite; 65 Munakataite, 65 Schmiederite, 65 Linarite; 70 Chenite, 75 Krivovichevite.
  - 07.BD With only large cations: 05 Sulphohalite; 10 Galeite, 10 Schairerite; 15 Kogarkoite; 20 Cesanite, 20 Caracolite; 25 Burkeite, 30 Hanksite, 35 Cannonite, 40 Lanarkite, 45 Grandreefite, 50 Itoite, 55 Chiluite, 60 Hectorfloresite, 65 Pseudograndreefite, 70 Sundiusite.
- 07.C Sulfates (selenates, etc.) without Additional Anions, with H_2O:
  - 07.CB With only medium-sized cations: 05 Gunningite, 05 Dwornikite, 05 Kieserite, 05 Szomolnokite, 05 Szmikite, 05 Poitevinite, 05 Cobaltkieserite; 07 Sanderite, 10 Bonattite, 15 Boyleite, 15 Aplowite, 15 Ilesite, 15 Rozenite, 15 Starkeyite, 15 IMA2002-034; 20 Chalcanthite, 20 Jokokuite, 20 Pentahydrite, 20 Siderotil; 25 Bianchite, 25 Ferrohexahydrite, 25 Chvaleticeite, 25 Hexahydrite, 25 Moorhouseite, 25 Nickelhexahydrite; 30 Retgersite; 35 Bieberite, 35 Boothite, 35 Mallardite, 35 Melanterite, 35 Zincmelanterite, 35 Alpersite; 40 Epsomite, 40 Goslarite, 40 Morenosite; 45 Alunogen, 45 Meta-alunogen; 50 Coquimbite, 50 Paracoquimbite; 55 Rhomboclase, 60 Kornelite, 65 Quenstedtite, 70 Lausenite; 75 Lishizhenite, 75 Romerite; 80 Ransomite; 85 Bilinite, 85 Apjohnite, 85 Dietrichite, 85 Halotrichite, 85 Pickeringite, 85 Redingtonite, 85 Wupatkiite; 90 Meridianiite, 95 Caichengyunite.
  - 07.CC With medium-sized and large cations: 05 Krausite, 10 Tamarugite; 15 Mendozite, 15 Kalinite; 20 Lonecreekite, 20 Alum-(K), 20 Alum-(Na), 20 Lanmuchangite, 20 Tschermigite; 25 Pertlikite, 25 Monsmedite?, 25 Voltaite, 25 Zincovoltaite; 30 Krohnkite, 35 Ferrinatrite, 40 Goldichite, 45 Loweite; 50 Blodite, 50 Changoite, 50 Nickelblodite; 55 Mereiterite, 55 Leonite; 60 Boussingaultite, 60 Cyanochroite, 60 Mohrite, 60 Picromerite, 60 Nickelboussingaultite; 65 Polyhalite; 70 Leightonite, 75 Amarillite, 80 Konyaite, 85 Wattevilleite.
  - 07.CD With only large cations: 05 Matteuccite, 10 Mirabilite, 15 Lecontite, 20 Hydroglauberite, 25 Eugsterite, 30 Gorgeyite; 35 Koktaite, 35 Syngenite; 40 Gypsum, 45 Bassanite, 50 Zircosulfate, 55 Schieffelinite, 60 Montanite, 65 Omongwaite
- 07.D Sulfates (selenates, etc.) with additional anions, with H_2O:
  - 07.DB With only medium-sized cations; insular octahedra and finite groups: 05 Svyazhinite, 05 Aubertite, 05 Magnesioaubertite; 10 Rostite, 10 Khademite; 15 Jurbanite; 20 Minasragrite, 20 Anorthominasragrite, 20 Orthominasragrite; 25 Bobjonesite; 30 Amarantite, 30 Hohmannite, 30 Metahohmannite; 35 Aluminocopiapite, 35 Copiapite, 35 Calciocopiapite, 35 Cuprocopiapite, 35 Ferricopiapite, 35 Magnesiocopiapite, 35 Zincocopiapite.

- 07.DC With only medium-sized cations; chains of corner-sharing octahedra: 05 Aluminite, 05 Meta-aluminite; 10 Butlerite, 10 Parabutlerite; 15 Fibroferrite, 20 Xitieshanite; 25 Botryogen, 25 Zincobotryogen; 30 Chaidamuite, 30 Guildite.
- 07.DD With only medium-sized cations; sheets of edge-sharing octahedra: 05 Basaluminite?, 05 Felsobanyaite, 07.5 Kyrgyzstanite, 08.0 Zn-Schulenbergite; 10 Langite, 10 Posnjakite, 10 Wroewolfeite; 15 Spangolite, 20 Ktenasite, 25 Christelite; 30 Campigliaite, 30 Devilline, 30 Orthoserpierite, 30 Niedermayrite, 30 Serpierite; 35 Motukoreaite, 35 Mountkeithite, 35 Glaucocerinite, 35 Honessite, 35 Hydrowoodwardite, 35 Hydrohonessite, 35 Shigaite, 35 Natroglaucocerinite, 35 Wermlandite, 35 Nikischerite, 35 Zincaluminite, 35 Woodwardite, 35 Carrboydite, 35 Zincowoodwardite, 35 Zincowoodwardite-3R, 35 Zincowoodwardite-1T; 40 Lawsonbauerite, 40 Torreyite, 45 Mooreite, 50 Namuwite, 55 Bechererite, 60 Ramsbeckite, 65 Vonbezingite, 70 Redgillite; 75 Chalcoalumite, 75 Nickelalumite*; 80 Guarinoite, 80 Theresemagnanite, 80 Schulenbergite; 85 Montetrisaite.
- 07.DE With only medium-sized cations; unclassified: 05 Mangazeite; 10 Carbonate-cyanotrichite, 10 Cyanotrichite; 15 Schwertmannite, 20 Tlalocite, 25 Utahite, 35 Coquandite, 40 Osakaite, 45 Wilcoxite, 50 Stanleyite, 55 Mcalpineite, 60 Hydrobasaluminite, 65 Zaherite, 70 Lautenthalite, 75 Camérolaite, 80 Brumadoite.
- 07.DF With large and medium-sized cations: 05 Uklonskovite, 10 Kainite, 15 Natrochalcite; 20 Metasideronatrite, 20 Sideronatrite; 25 Despujolsite, 25 Fleischerite, 25 Schaurteite, 25 Mallestigite; 30 Slavikite, 35 Metavoltine; 40 Lannonite, 40 Vlodavetsite; 45 Peretaite, 50 Gordaite, 55 Clairite, 60 Arzrunite, 65 Elyite, 70 Yecoraite, 75 Riomarinaite, 80 Dukeite, 85 Xocolatlite.
- 07.DG With large and medium-sized cations; with NO_3, CO_3, $B(OH)_4$, SiO_4 or IO_3: 05 Darapskite; 10 Clinoungemachite, 10 Ungemachite, 10 Humberstonite; 15 Bentorite, 15 Charlesite, 15 Ettringite, 15 Jouravskite, 15 Sturmanite, 15 Thaumasite, 15 Carraraite, 15 Buryatite; 20 Rapidcreekite, 25 Tatarskite, 30 Nakauriite, 35 Chessexite; 40 Carlosruizite, 40 Fuenzalidaite; 45 Chelyabinskite*.

- 07.E Uranyl Sulfates:
  - 07.EA Without cations: 05 Uranopilite, 05 Metauranopilite, 10 Jachymovite.
  - 07.EB With medium-sized cations: 05 Johannite, 10 Deliensite.
  - 07.EC With medium-sized and large cations: 05 Cobaltzippeite, 05 Magnesiozippeite, 05 Nickelzippeite, 05 Natrozippeite, 05 Zinc-zippeite, 05 Zippeite; 10 Rabejacite, 15 Marecottite, 20 Pseudojohannite.
- 07.J Thiosulfates:
  - 07.JA Thiosulfates with Pb: 05 Sidpietersite.
- 07.X Unclassified Strunz Sulfates (Selenates, Tellurates):
  - 07.XX Unknown: 00 Aiolosite, 00 Steverustite, 00 Grandviewite, 00 IMA2009-008, 00 Adranosite, 00 Blakeite.

Class: Chromates

- 07.F Chromates:
    - 07.FA Without additional anions: 05 Tarapacaite, 10 Chromatite, 15 Hashemite, 20 Crocoite.
    - 07.FB With additional O, V, S, Cl: 05 Phoenicochroite, 10 Santanaite, 15 Wattersite, 20 Deanesmithite, 25 Edoylerite.
    - 07.FC With PO_4, AsO_4, SiO_4: 05 Vauquelinite; 10 Fornacite, 10 Molybdofornacite; 15 Hemihedrite, 15 Iranite; 20 Embreyite, 20 Cassedanneite.
    - 07.FD Dichromates: 05 Lopezite.

Class: Molybdates, Wolframates And Niobates

- 07.G Molybdates, wolframates and niobates:
    - 07.GA Without additional anions or H_2O: 05 Fergusonite-(Ce), 05 Fergusonite-(Nd) ^{N}, 05 Fergusonite-(Y), 05 Powellite, 05 Wulfenite, 05 Stolzite, 05 Scheelite; 10 Formanite-(Y), 10 Iwashiroite-(Y); 15 Paraniite-(Y).
    - 07.GB With additional anions and/or H_2O: 05 Lindgrenite, 10 Szenicsite, 15 Cuprotungstite, 20 Phyllotungstite, 25 Rankachite, 30 Ferrimolybdite, 35 Anthoinite, 35 Mpororoite, 40 Obradovicite-KCu, 45 Mendozavilite-NaFe, 45 Paramendozavilite, 50 Tancaite-(Ce).
- 07.H Uranium and uranyl molybdates and wolframates:
    - 07.HA With U^{4+}: 05 Sedovite, 10 Cousinite, 15 Moluranite.
    - 07.HB With U^{6+}: 15 Calcurmolite, 20 Tengchongite, 25 Uranotungstite.

Phosphate Minerals

Phosphate minerals are those minerals that contain the tetrahedrally coordinated phosphate (PO_4^{3-}) anion along with the freely substituting arsenate (AsO_4^{3-}) and vanadate (VO_4^{3-}). Chlorine (Cl^-), fluorine (F^-), and hydroxide (OH^-) anions also fit into the crystal structure.

The phosphate class of minerals is a large and diverse group, however, only a few species are relatively common.

Examples

Phosphate minerals include:

- Triphylite $Li(Fe,Mn)PO_4$.
- Monazite $(Ce,La,Y,Th)PO_4$.
- Hinsdalite $PbAl_3(PO_4)(SO_4)(OH)_6$.

- Pyromorphite $Pb_5(PO_4)_3Cl$.
- Vanadinite $Pb_5(VO_4)_3Cl$.
- Erythrite $Co_3(AsO_4)_2·8H_2O$.
- Amblygonite $LiAlPO_4F$.
- Lazulite $(Mg,Fe)Al_2(PO_4)_2(OH)_2$.
- Wavellite $Al_3(PO_4)_2(OH)_3·5H_2O$.
- Turquoise $CuAl_6(PO_4)_4(OH)_8·5H_2O$.
- Autunite $Ca(UO_2)_2(PO_4)_2·10\text{-}12H_2O$.
- Carnotite $K_2(UO_2)_2(VO_4)_2·3H_2O$.
- Phosphophyllite $Zn_2(Fe,Mn)(PO_4)_2•4H_2O$.
- Struvite $(NH_4)MgPO_4·6H_2O$.
- Xenotime-Y $Y(PO_4)$.
- Apatite group $Ca_5(PO_4)_3(F,Cl,OH)$.
    - Hydroxylapatite $Ca_5(PO_4)_3OH$.
    - Fluorapatite $Ca_5(PO_4)_3F$.
    - Chlorapatite $Ca_5(PO_4)_3Cl$.
    - Bromapatite.
- Mitridatite group:
    - Arseniosiderite-mitridatite series $(Ca_2(Fe^{3+})_3[(O)_2|(AsO_4)_3]·3H_2O$ -- $Ca_2(Fe^{3+})_3[(O)_2|(PO_4)_3]·3H_2O)$
    - Arseniosiderite-robertsite series $(Ca_2(Fe^{3+})_3[(O)_2|(AsO_4)_3]·3H_2O$ -- $Ca_3(Mn^{3+})_4[(OH)_3|(PO_4)_2]_2·3H_2O)$

Applications

Phosphate rock is a general term that refers to rock with high concentration of phosphate minerals, most commonly of the apatite group. It is the major resource mined to produce phosphate fertilizers for the agriculture sector. Phosphate is also used in animal feed supplements, food preservatives, anti-corrosion agents, cosmetics, fungicides, ceramics, water treatment and metallurgy.

The largest use of minerals mined for their phosphate content is the production of fertilizer.

Phosphate minerals are often used for control of rust and prevention of corrosion on ferrous materials applied with electrochemical conversion coatings.

Nickel–Strunz Classification -08- Phosphates

IMA-CNMNC proposes a new hierarchical scheme (Mills et al., 2009). This list uses it to modify the Classification of Nickel–Strunz (mindat.org, 10 ed, pending publication).

- Abbreviations:
    - "*" - discredited (IMA/CNMNC status).
    - "?" - questionable/doubtful (IMA/CNMNC status).
    - "REE" - Rare-earth element (Sc, Y, La, Ce, Pr, Nd, Pm, Sm, Eu, Gd, Tb, Dy, Ho, Er, Tm, Yb, Lu).
    - "PGE" - Platinum-group element (Ru, Rh, Pd, Os, Ir, Pt).
    - 03.C Aluminofluorides, 06 Borates, 08 Vanadates (04.H $V^{[5,6]}$ Vanadates), 09 Silicates:
        - Neso: insular.
        - Soro: grouping.
        - Cyclo: ring.
        - Ino: chain.
        - Phyllo: sheet.
        - Tekto: three-dimensional framework.
- Nickel–Strunz code scheme: NN.XY.##x:
    - NN: Nickel–Strunz mineral class number.
    - X: Nickel–Strunz mineral division letter.
    - Y: Nickel–Strunz mineral family letter.
    - ##x: Nickel–Strunz mineral/group number, x add-on letter.

Class: Phosphates

- 08.A Phosphates, etc. without additional anions, without H_2O:
    - 08.AA With small cations (some also with larger ones): 05 Berlinite, 05 Rodolicoite; 10 Beryllonite, 15 Hurlbutite, 20 Lithiophosphate, 25 Nalipoite, 30 Olympite.
    - 08.AB With medium-sized cations: 05 Farringtonite; 10 Ferrisicklerite, 10 Heterosite, 10 Natrophilite, 10 Lithiophilite, 10 Purpurite, 10 Sicklerite, 10 Simferite, 10 Triphylite; 15 Chopinite, 15 Sarcopside; 20 Beusite, 20 Graftonite.
    - 08.AC With medium-sized and large cations: 10 IMA2008-054, 10 Alluaudite, 10 Hagendorfite, 10 Ferroalluaudite, 10 Maghagendorfite, 10 Varulite, 10 Ferrohagendorfite*; 15 Bobfergusonite, 15 Ferrowyllieite, 15 Qingheiite, 15 Rosemaryite, 15 Wyllieite, 15 Ferrorosemaryite; 20 Maricite, 30 Brianite, 35 Vitusite-(Ce); 40 Olgite?, 40 Bario-olgite; 45 Ferromerrillite, 45 Bobdownsite, 45 Merrillite-(Ca)*, 45 Merrillite, 45

Merrillite-(Y)*, 45 Whitlockite, 45 Tuite, 45 Strontiowhitlockite; 50 Stornesite-(Y), 50 Xenophyllite, 50 Fillowite, 50 Chladniite, 50 Johnsomervilleite, 50 Galileiite; 55 Harrisonite, 60 Kosnarite, 65 Panethite, 70 Stanfieldite, 90 IMA2008-064.

- 08.AD With only large cations: 05 Nahpoite, 10 Monetite, 15 Archerite, 15 Biphosphammite; 20 Phosphammite, 25 Buchwaldite; 35 Pretulite, 35 Xenotime-(Y), 35 Xenotime-(Yb); 45 Ximengite, 50 Monazite-(Ce), 50 Monazite-(La), 50 Monazite-(Nd), 50 Monazite-(Sm), 50 Brabantite?

- 08.B Phosphates, etc. with Additional Anions, without H_2O:
  - 08.BA With small and medium-sized cations: 05 Vayrynenite; 10 Hydroxylherderite, 10 Herderite; 15 Babefphite.
  - 08.BB With only medium-sized cations, (OH, etc.):RO_4 £1:1: 05 Amblygonite, 05 Natromontebrasite?, 05 Montebrasite?, 05 Tavorite; 10 Zwieselite, 10 Triplite, 10 Magniotriplite?, 10 Hydroxylwagnerite; 15 Joosteite, 15 Stanekite, 15 Triploidite, 15 Wolfeite, 15 Wagnerite; 20 Satterlyite, 20 Holtedahlite; 25 Althausite; 30 Libethenite, 30 Zincolibethenite; 35 Tarbuttite; 40 Barbosalite, 40 Hentschelite, 40 Scorzalite, 40 Lazulite; 45 Trolleite, 55 Phosphoellenbergerite; 90 Zinclipscombite, 90 Lipscombite, 90 Richellite.
  - 08.BC With only medium-sized cations, (OH, etc.):RO_4 > 1:1 and < 2:1: 10 Plimerite, 10 Frondelite, 10 Rockbridgeite.
  - 08.BD With only medium-sized cations, (OH, etc.):RO_4 = 2:1: 05 Pseudomalachite, 05 Reichenbachite, 10 Gatehouseite, 25 Ludjibaite.
  - 08.BE With only medium-sized cations, (OH, etc.):RO_4 > 2:1: 05 Augelite, 10 Grattarolaite, 15 Cornetite, 30 Raadeite, 85 Waterhouseite.
  - 08.BF With medium-sized and large cations, (OH, etc.):RO_4 < 0.5:1: 05 Arrojadite, 05 Arrojadite-(BaFe), 05 Arrojadite-(KFe), 05 Arrojadite-(NaFe), 05 Arrojadite-(SrFe), 05 Arrojadite-(KNa), 05 Arrojadite-(PbFe), 05 Arrojadite-(BaNa), 05 Fluorarrojadite-(BaNa), 05 Fluorarrojadite-(KNa), 05 Fluorarrojadite-(BaFe), 05 Ferri-arrojadite-(BaNa), 05 Dickinsonite, 05 Dickinsonite-(KNa), 05 Dickinsonite-(KMnNa), 05 Dickinsonite-(KNaNa), 05 Dickinsonite-(NaNa); 10 Samuelsonite, 15 Griphite, 20 Nabiasite.
  - 08.BG With medium-sized and large cations, (OH, etc.):RO_4 = 0.5:1: 05 Bearthite, 05 Goedkenite, 05 Tsumebite; 10 Melonjosephite, 15 Tancoite.
  - 08.BH With medium-sized and large cations, (OH,etc.):RO_4 = 1:1: 05 Thadeuite; 10 Lacroixite, 10 Isokite, 10 Panasqueiraite; 15 Drugmanite; 20 Bjarebyite, 20 Kulanite, 20 Penikisite, 20 Perloffite, 20 Johntomaite; 25 Bertossaite, 25 Palermoite; 55 Jagowerite, 60 Attakolite.
  - 08.BK With medium-sized and large cations, (OH, etc.): 05 Brazilianite, 15 Curetonite, 25 Lulzacite.
  - 08.BL With medium-sized and large cations, (OH, etc.):RO_4 = 3:1: 05 Corkite, 05 Hinsdalite, 05 Orpheite, 05 Woodhouseite, 05 Svanbergite; 10 Kintoreite, 10 Benauite, 10 Crandallite, 10 Goyazite, 10 Springcreekite, 10 Gorceixite; 10 Lusungite?, 10

Plumbogummite, 10 Ferrazite?; 13 Eylettersite, 13 Florencite-(Ce), 13 Florencite-(La), 13 Florencite-(Nd), 13 Waylandite, 13 Zairite; 15 Viitaniemiite, 20 Kuksite, 25 Pattersonite.

- 08.BM With medium-sized and large cations, (OH, etc.):RO_4 = 4:1: 10 Paulkellerite, 15 Brendelite.
- 08.BN With only large cations, (OH, etc.):RO_4 = 0.33:1: 05 IMA2008-068, 05 Phosphohedyphane, 05 IMA2008-009, 05 Alforsite, 05 Apatite*, 05 Apatite-(CaOH), 05 Apatite-(CaCl), 05 Apatite-(CaF), 05 Apatite-(SrOH), 05 Apatite-(CaOH)-M, Carbonate-fluorapatite?, 05 Carbonate-hydroxylapatite?, 05 Belovite-(Ce), 05 Belovite-(La), 05 Fluorcaphite, 05 Pyromorphite, 05 Hydroxylpyromorphite, 05 Deloneite-(Ce), 05 Kuannersuite-(Ce), 10 Arctite.
- 08.BO With only large cations, (OH, etc.):RO_4 1:1: 05 Nacaphite, 10 Petitjeanite, 15 Smrkovecite, 25 Heneuite, 30 Nefedovite, 40 Artsmithite.

- 08.C Phosphates without Additional Anions, with H_2O:
  - 08.CA With small and large/medium cations: 05 Fransoletite, 05 Parafransoletite; 10 Ehrleite, 15 Faheyite; 20 Gainesite, 20 Mccrillisite, 20 Selwynite; 25 Pahasapaite, 30 Hopeite, 40 Phosphophyllite; 45 Parascholzite, 45 Scholzite; 65 Gengenbachite, 70 Parahopeite.
  - 08.CB With only medium-sized cations, RO_4:H_2O = 1:1: 05 Serrabrancaite, 10 Hureaulite.
  - 08.CC With only medium-sized cations, RO_4:H_2O = 1:1.5: 05 Garyansellite, 05 Kryzhanovskite, 05 Landesite, 05 Phosphoferrite, 05 Reddingite.
  - 08.CD With only medium-sized cations, RO_4:H_2O = 1:2: 05 Kolbeckite, 05 Metavariscite, 05 Phosphosiderite; 10 Strengite, 10 Variscite; 20 Ludlamite.
  - 08.CE With only medium-sized cations, RO_4:H_2O £1:2.5: 10 Newberyite, 20 Phosphorrosslerite; 25 Metaswitzerite, 25 Switzerite; 35 Bobierrite; 40 Arupite, 40 Baricite, 40 Vivianite, 40 Pakhomovskyite; 50 Cattiite, 55 Koninckite; 75 IMA2008-046, 75 Malhmoodite; 80 Santabarbaraite, 85 Metavivianite.
  - 08.CF With large and medium-sized cations, RO_4:H_2O > 1:1: 05 Tassieite, 05 Wicksite, 05 Bederite; 10 Haigerachite.
  - 08.CG With large and medium-sized cations, RO_4:H_2O = 1:1: 05 Collinsite, 05 Cassidyite, 05 Fairfieldite, 05 Messelite, 05 Hillite, (05 Uranophane-beta but Uranophane 09.AK.15); 20 Phosphogartrellite.
  - 08.CH With large and medium-sized cations, RO_4:H_2O < 1:1: 10 Anapaite, 20 Dittmarite, 20 Niahite, 25 Francoanellite, 25 Taranakite, 30 Schertelite, 35 Hannayite, 40 Hazenite, 40 Struvite, 40 Struvite-(K), 45 Rimkorolgite, 50 Bakhchisaraitsevite, 55 IMA2008-048.
  - 08.CJ With only large cations: 05 Stercorite, 10 Mundrabillaite, 10 Swaknoite, 15 Nastrophite, 15 Nabaphite, 45 Brockite, 45 Grayite, 45 Rhabdophane-(Ce), 45 Rhabdophane-(La), 45 Rhabdophane-(Nd), 45 Tristramite, 50 Brushite, 50 Churchite-(Dy)*, 50 Churchite-(Nd), 50 Churchite-(Y), 50 Ardealite, 60 Dorfmanite, 70 Catalanoite, 80 Ningyoite.

- 08.D Phosphates:
  - 08.DA With small (and occasionally larger) cations: 05 Moraesite, 10 Footemineite, 10 Ruifrancoite, 10 Guimaraesite, 10 Roscherite, 10 Zanazziite, 10 Atencioite, 10 Greifensteinite; 15 Uralolite, 20 Weinebeneite, 25 Tiptopite, 30 Veszelyite, 35 Kipushite, 40 Spencerite, 45 Glucine.
  - 08.DB With only medium-sized cations, (OH, etc.):RO_4 < 1:1: 05 Diadochite, 10 Vashegyite, 15 Schoonerite, 20 Sinkankasite, 25 Mitryaevaite, 30 Sanjuanite, 50 Giniite, 55 Sasaite, 60 Mcauslanite, 65 Goldquarryite, 70 Birchite.
  - 08.DC With only medium-sized cations, (OH, etc.):RO_4 = 1:1 and < 2:1: 05 Nissonite; 15 Kunatite, 15 Earlshannonite, 15 Whitmoreite; 17 Kleemanite, 20 Bermanite, 20?? Oxiberaunite*, 22 Kovdorskite; 25 Ferrostrunzite, 25 Ferristrunzite, 25 Metavauxite, 25 Strunzite; 27 Beraunite; 30 Gordonite, 30 Laueite, 30 Sigloite, 30 Paravauxite, 30 Ushkovite, 30 Ferrolaueite, 30 Mangangordonite, 30 Pseudolaueite, 30 Stewartite, 30 Kastningite, 35 Vauxite, 37 Vantasselite, 40 Cacoxenite; 45 Gormanite, 45 Souzalite; 47 Kingite; 50 Wavellite, 50 Allanpringite, 52 Kribergite, 60 Nevadaite.
  - 08.DD With only medium-sized cations, (OH, etc.):RO_4 = 2:1: 15 Aheylite, 15 Chalcosiderite, 15 Faustite, 15 Planerite, 15 Turquoise; 20 Ernstite, 20 Childrenite, 20 Eosphorite.
  - 08.DE With only medium-sized cations, (OH, etc.):RO_4 = 3:1: 05 Senegalite, 10 Fluellite, 20 Zapatalite, (35 Alumoakermanite, Mindat.org: 09.BB.10), 35 Aldermanite.
  - 08.DF With only medium-sized cations, (OH,etc.):RO_4 > 3:1: 05 Hotsonite-VII, 05 Hotsonite-VI; 10 Bolivarite, 10 Evansite, 10 Rosieresite, 25 Sieleckiite, 40 Gladiusite.
  - 08.DG With large and medium-sized cations, (OH, etc.):RO_4 < 0.5:1: 05 Sampleite.
  - 08.DH With large and medium-sized cations, (OH, etc.):RO_4 < 1:1: 05 Minyulite; 10 Leucophosphite, 10 Spheniscidite, 10 Tinsleyite; 15 Kaluginite*, 15 Keckite, 15 Jahnsite-(CaMnFe), 15 Jahnsite-(CaMnMg), 15 Jahnsite-(CaMnMn), 15 Jahnsite-(MnMnMn)*, 15 Jahnsite-(CaFeFe), 15 Jahnsite-(NaFeMg), 15 Jahnsite-(CaMgMg), 15 Jahnsite-(NaMnMg), 15 Rittmannite, 15 Whiteite-(MnFeMg), 15 Whiteite-(CaFeMg), 15 Whiteite-(CaMnMg); 20 Manganosegelerite, 20 Overite, 20 Segelerite, 20 Wilhelmvierlingite, 20 Juonniite; 25 Calcioferrite, 25 Kingsmountite, 25 Montgomeryite, 25 Zodacite; 30 Lunokite, 30 Pararobertsite, 30 Robertsite, 30 Mitridatite; 35 Matveevite?, 35 Mantienneite, 35 Paulkerrite, 35 Benyacarite, 40 Xanthoxenite, 55 Englishite.
  - 08.DJ With large and medium-sized cations, (OH, etc.):RO_4 = 1:l: 05 Johnwalkite, 05 Olmsteadite, 10 Gatumbaite, 20 Meurigite-Na, 20 Meurigite-K, 20 Phosphofibrite, 25 Jungite, 30 Wycheproofite, 35 Ercitite, 40 Mrazekite.
  - 08.DK With large and medium-sized cations, (OH, etc.):RO_4 > 1:1 and < 2:1: 15 Matioliite, 15 IMA2008-056, 15 Dufrenite, 15 Burangaite, 15 Natrodufrenite; 20 Kidwellite, 25 Bleasdaleite, 30 Matulaite, 35 Krasnovite.

- 08.DL With large and medium-sized cations, (OH, etc.):RO_4 = 2:1: 05 Foggite; 10 Cyrilovite, 10 Millisite, 10 Wardite; 15 Petersite-(Y), 15 Calciopetersite; 25 Angastonite.
- 08.DM With large and medium-sized cations, (OH, etc.):RO_4 > 2:1: 05 Morinite, 15 Melkovite, 25 Gutsevichite?, 35 Delvauxite.
- 08.DN With only large cations: 05 Natrophosphate, 10 Isoclasite, 15 Lermontovite, 20 Vyacheslavite.
- 08.DO With CO_3, SO_4, SiO_4: 05 Girvasite, 10 Voggite, 15 Peisleyite, 20 Perhamite, 25 Saryarkite-(Y), 30 Micheelsenite, 40 Parwanite, 45 Skorpionite.

- 08.E Uranyl Phosphates:
    - 08.EA UO_2:RO_4 = 1:2: 05 Phosphowalpurgite, 10 Parsonsite, 15 Ulrichite, 20 Lakebogaite.
    - 08.EB UO_2:RO_4 = 1:1: 05 Autunite, 05 Uranocircite, 05 Torbernite, 05 Xiangjiangite, 05 Saleeite; 10 Bassetite, 10 Meta-autunite, 10 Metauranocircite, 10 Metatorbernite, 10 Lehnerite, 10 Przhevalskite; 15 Chernikovite, 15 Meta-ankoleite, 15 Uramphite; 20 Threadgoldite, 25 Uranospathite, 30 Vochtenite, 35 Coconinoite, 40 Ranunculite, 45 Triangulite, 50 Furongite, 55 Sabugalite.
    - 08.EC UO_2:RO_4 = 3:2: 05 Francoisite-(Ce), 05 Francoisite-(Nd), 05 Phuralumite, 05 Upalite; 10 Kivuite?, 10 Yingjiangite, 10 Renardite, 10 Dewindtite, 10 Phosphuranylite; 15 Dumontite; 20 Metavanmeersscheite, 20 Vanmeersscheite; 25 Althupite, 30 Mundite, 35 Phurcalite, 40 Bergenite.
    - 08.ED Unclassified: 05 Moreauite, 10 Sreinite, 15 Kamitugaite.
- 08.F Polyphosphates:
    - 08.FA Polyphosphates, without OH and H_2O; dimers of corner-sharing RO_4 tetrahedra: 20 Pyrocoproite*, 20 Pyrophosphite*.
    - 08.FC Polyphosphates, with H_2O only: 10 Canaphite, 20 Arnhemite*, 25 Wooldridgeite, 30 Kanonerovite.
- 08.X Unclassified Strunz Phosphates:
    - 08.XX Unknown: 00 Sodium-autunite, 00 Pseudo-autunite*, 00 Cheralite-(Ce)?, 00 Laubmannite?, 00 Spodiosite?, 00 Sodium meta-autunite, 00 Kerstenite?, 00 Lewisite, 00 Coeruleolactite, 00 Viseite, 00 IMA2009-005.

References

- Ernest H. Nickel and Monte C. Nichols (2008-05-22). "IMA/CNMNC List of Mineral Name based on the database MINERAL, which Materials Data, Inc. (MDI) makes available" (PDF). Retrieved 17, April 2020
- Knobloch, Eberhard (2003). The shoulders on which We stand/Wegbereiter der Wissenschaft (in German and English). Springer. pp. 170–173. ISBN 3-540-20557-8
- Bindi, L.; Paul J. Steinhardt; Nan Yao; Peter J. Lu (2011). "Icosahedrite, $Al_{63}Cu_{24}Fe_{13}$, the first natural quasicrystal" (PDF). American Mineralogist. 96: 928–931. doi:10.2138/am.2011.3758

- Dyar, Gunter, and Tasa (2007). Mineralogy and Optical Mineralogy. Mineralogical Society of America. pp. 2–4. ISBN 978-0939950812
- Hurlbut, Cornelius S.; Klein, Cornelis, Manual of Mineralogy, 20th ed., John Wiley and Sons, New York ISBN 0-471-80580-7

Chapter 5

Gemstone Minerals

Gemstones are minerals which are highly prized for their beauty, durability and rarity. Gemstones, after being cut and polished, can be used in jewelry. Gemology is the science of studying, cutting and valuing of precious stones regardless of whether they are artificial, imitation or natural This chapter focuses on gemstones and gemology and will help in obtaining extensive understanding of the subject matter.

Gemology is the combined art and science of studying, cutting, valuing, buying and selling precious stones. Some of the most precious jewels that gemologists deal in include diamonds, rubies, sapphires, and emeralds. Gemology is a type of mineralogy that specifically focuses on the study of precious gems, both natural and those composed of artificial materials. Some jewelers are gemologists. Appraisers who evaluate the value of gemstones should also be a gemologist or at least have studied extensively to master the most important aspects of determining a gem's condition and value.

With advances in gemstone synthesis, gemology has become a much more important field of study. Often, industry professionals are the most highly trained and educated gemologists. These professionals have become certified gemologists. An accredited gemologist is an expert in the field who has completed an intense training and gone through advanced education in various gemstone areas. Often, they will specialize in a specific area of the business or a particular type of stone. Many gemologists specialize in the study and evaluation of diamonds.

Gemologists value stones based on factors, including cut, color, quality, and clarity. Many people are familiar with a similar group of criteria for gemstone evaluation used in assessing diamonds. For diamonds, the factors are generally known as the 4 Cs of color, clarity, cut and carat.

Gemology and Gemstone Investing

When returns in the stock market decline, aggressive investors often seek out alternative investments which hold more promise of increasing returns on invested capital. Precious gems, in particular, those that are rare or of exceptional quality, have a high likelihood of at least retaining their value, and are likely to increase in value. However, they are not an asset which may be as easily liquidated as other investments if there is an urgent need for cash. This problem is especially real for rare, precious stones which would appeal to an elite level of buyer.

Gemstone investing can seem exciting to people who want to make quick returns, but it is highly speculative and should only be undertaken by experienced professionals. The term investment-grade is often tossed around by those who want to sell gems or try to convince other people to invest in them. However, this is controversial and frowned upon by industry leaders, as there are no formal standards for what constitutes an investment-grade gemstone.

Gemstone

A gemstone is the name given to a piece of mineral (or other rock or organic material) that after it has been cut and polished has been made into a piece of jewellery or another accessory. Although gemstones are commonly made from minerals, material such as jet or amber or rocks like lapis lazuli can also be utilised in the creation of gemstones. The majority of gemstones will be hard to the touch but depending on the piece of jewellery that is being created, some soft minerals are also utilised. Take a look at Moh's Scale of Hardness to find out more about this.

Some other names for a gemstone include:

- Gem/Fine Gem,
- Precious stone/Semi-precious stone,
- Jewel.

Formation of Gemstones

Even though the vast majority of gemstones are created below the surface of the earth, there are a number of different ways that the individual stones can be created. A great number of these precious stones are created when the minerals react with water that is found just below the surface of the earth and the minerals then dissolve. This process allows the minerals to form with gemstones like opals, agates and amethysts forming when the solution cools or evaporates.

Gems like malachite, azurite or turquoise are formed when water blends with rocks that are rich in copper. There is a strong history attached to turquoise gems with the ancient Egyptians regarding them to be an extremely precious gem. This means that turquoise gems have been around for 4,000 years or even longer.

Value of Gemstones

Much the same as you would expect from any other major industry, the price of gemstones is largely influenced by supply and demand.

One of the most important things for people when considering gemstones is the value and price for them but like in many industries, the laws of supply and demand are very much in place. This means that rare gems will carry a higher price tag.

For instance, a garnet is extremely stylish and attractive but it is not that rare, which means that its price tag is not too high. If you compared a garnet to a ruby of a similar size and condition, the ruby would always hold a higher price, simply because of the fact that it is a rare stone.

It is often difficult to place a true value on any particular gemstone because there is no uniform classification or measurement system in place. This is where gemstones differ from diamonds, which have been subject to a uniform measurement system since the 1950s. Traditionally, gemstones have been examined and evaluated by eye. When examining a gemstone, the four most important factors can be summed up in the four C's, which are; colour, clarity, cut and carat.

Even though diamonds are one of the most prized possessions on Earth, there are actually a number of different gemstones that are rarer to find than diamonds. Some of the rarest gemstones you can find include:

- Painite,
- Alexandrite,
- Tanzanite,
- Benitoite,
- Poudretteite,
- Grandiderite,
- Musgravite,
- Jeremejevite,
- Red Beryl.

Cutting and Polishing

A few gemstones are used as gems in the crystal or other form in which they are found. Most however, are cut and polished for usage as jewelry. The picture to the left is of a rural, commercial cutting operation in Thailand. This small factory cuts thousands of carats of sapphire annually. The two main classifications are stones cut as smooth, dome shaped stones called cabochons, and stones which are cut with a faceting machine by polishing small flat windows called facets at regular intervals at exact angles.

Raw gemstones

A diamond cutter in Amsterdam in the Netherlands in 2012.

Stones which are opaque or semi-opaque such as opal, turquoise, variscite, etc. are commonly cut as cabochons. These gems are designed to show the stone's color or surface properties as in opal and star sapphires. Grinding wheels and polishing agents are used to grind, shape and polish the smooth dome shape of the stones.

Gems which are transparent are normally faceted, a method which shows the optical properties of the stone's interior to its best advantage by maximizing reflected light which is perceived by the viewer as sparkle. There are many commonly used shapes for faceted stones. The facets must be cut at the proper angles, which varies depending on the optical properties of the gem. If the angles are too steep or too shallow, the light will pass through and not be reflected back toward the viewer. The faceting machine is used to hold the stone onto a flat lap for cutting and polishing the flat facets. Rarely, some cutters use special curved laps to cut and polish curved facets.

Colors

Nearly 300 variations of diamond color exhibited at the Aurora display at the Natural History Museum in London.

The color of any material is due to the nature of light itself. Daylight, often called white light, is actually all of the colors of the spectrum combined. When light strikes a material, most of the light is absorbed while a smaller amount of a particular frequency or wavelength is reflected. The part that is reflected reaches the eye as the perceived color. A ruby appears red because it absorbs all the other colors of white light (green and blue), while reflecting the red.

A material which is mostly the same can exhibit different colors. For example, ruby and sapphire have the same primary chemical composition (both are corundum) but exhibit different colors because of impurities. Even the same named gemstone can occur in many different colors: sapphires show different shades of blue and pink and "fancy sapphires" exhibit a whole range of other colors from yellow to orange-pink, the latter called "padparadscha sapphire".

This difference in color is based on the atomic structure of the stone. Although the different stones formally have the same chemical composition and structure, they are not exactly the same. Every now and then an atom is replaced by a completely different atom, sometimes as few as one in a million atoms. These so-called impurities are sufficient to absorb certain colors and leave the other colors unaffected.

For example, beryl, which is colorless in its pure mineral form, becomes emerald with chromium impurities. If manganese is added instead of chromium, beryl becomes pink morganite. With iron, it becomes aquamarine.

Some gemstone treatments make use of the fact that these impurities can be “manipulated”, thus changing the color of the gem.

Treatment

Gemstones are often treated to enhance the color or clarity of the stone. Depending on the type and extent of treatment, they can affect the value of the stone. Some treatments are used widely because the resulting gem is stable, while others are not accepted most commonly because the gem color is unstable and may revert to the original tone.

Heat

Heat can improve gemstone color or clarity. The heating process has been well known to gem miners and cutters for centuries, and in many stone types heating is a common practice. Most citrine is made by heating amethyst, and partial heating with a strong gradient results in “ametrine” – a stone partly amethyst and partly citrine. Aquamarine is often heated to remove yellow tones, or to change green colors into the more desirable blue, or enhance its existing blue color to a purer blue.

Nearly all tanzanite is heated at low temperatures to remove brown undertones and give a more desirable blue/purple color. A considerable portion of all sapphire and ruby is treated with a variety of heat treatments to improve both color and clarity.

When jewelry containing diamonds is heated (for repairs) the diamond should be protected with boric acid; otherwise the diamond (which is pure carbon) could be burned on the surface or even burned completely up. When jewelry containing sapphires or rubies is heated, those stones should not be coated with boracic acid (which can etch the surface) or any other substance. They do not have to be protected from burning, like a diamond (although the stones *do* need to be protected from heat stress fracture by immersing the part of the jewelry with stones in water when metal parts are heated).

Radiation

Virtually all blue topaz, both the lighter and the darker blue shades such as “London” blue, has been irradiated to change the color from white to blue. Most greened quartz (Oro Verde) is also irradiated to achieve the yellow-green color. Diamonds are irradiated to produce fancy-color diamonds (which occur naturally, though rarely in gem quality).

Waxing/oiling

Emeralds containing natural fissures are sometimes filled with wax or oil to disguise them. This wax or oil is also colored to make the emerald appear of better color as well as clarity. Turquoise is also commonly treated in a similar manner.

Fracture Filling

Fracture filling has been in use with different gemstones such as diamonds, emeralds and sapphires. In 2006 "glass filled rubies" received publicity. Rubies over 10 carats (2 g) with large fractures were filled with lead glass, thus dramatically improving the appearance (of larger rubies in particular). Such treatments are fairly easy to detect.

Synthetic and Artificial Gemstones

It is important to distinguish between synthetic gemstones, and imitation or simulated gems.

Synthetic gems are physically, optically and chemically identical to the natural stone, but are created in controlled conditions in a laboratory. Imitation or simulated stones are chemically different than the natural stone but may be optically similar to it; they can be glass, plastic, resins or other compounds.

Examples of simulated or imitation stones include cubic zirconia, composed of zirconium oxide and simulated moissanite, which are both diamond simulants. The imitations copy the look and color of the real stone but possess neither their chemical nor physical characteristics. Moissanite actually has a higher refractive index than diamond and when presented beside an equivalently sized and cut diamond will have more "fire" than the diamond.

Synthetic, cultured or lab-created gemstones are not imitations. For example, diamonds, rubies, sapphires and emeralds have been manufactured in labs to possess identical chemical and physical characteristics to the naturally occurring variety. Synthetic (lab created) corundum, including ruby and sapphire, is very common and costs much less than the natural stones. Smaller synthetic diamonds have been manufactured in large quantities as industrial abrasives, although larger gem-quality synthetic diamonds are becoming available in multiple carats.

Whether a gemstone is a natural stone or lab-created (synthetic), the physical characteristics are the same. Lab-created stones tend to have a more vivid color to them, as impurities are not present in a lab and do not modify the clarity or color of the stone, unless added intentionally for a specific purpose.

Cardinal Gem

The term cardinal gemstone was traditionally used to refer to gemstones that were valued above all others. The five gemstones referred to as the cardinal gems included diamond, sapphire, ruby, emerald, and amethyst. The important use of these gemstones in traditional religious and cultural ceremonies and other events aided in labeling them as the cardinal gems. However, with the discovery of newer gemstones and new deposits of the cardinal gems in recent times, the term cardinal gems has largely fallen out of use.

Diamond (Gemstone)

Diamond is a mineral composed of pure carbon. It is the hardest naturally occurring substance known; it is also the most popular gemstone. Because of their extreme hardness, diamonds have a number of important industrial applications.

Diamond (Gem)

country	mine production 2006 (carats)*	% of world mine production
*Estimate.		
**Detail does not add to total given because of rounding.		
Australia	25,000,000	29.4
Congo (Kinshasa)	24,000,000	28.2
Russia	15,000,000	17.6
South Africa	9,000,000	10.6
Botswana	8,000,000	9.4
China	1,000,000	1.2
United States	300,000	0.4
Other countries	3,000,000	3.5
World total	85,000,000	100**

The hardness, brilliance, and sparkle of diamonds make them unsurpassed as gems. In the symbolism of gemstones, the diamond represents steadfast love and is the birthstone for April. Diamond stones are weighed in carats (1 carat = 200 milligrams) and in points (1 point = 0.01 carat). In addition to gem-quality stones, several varieties of industrial diamonds occur, and syntheticdiamonds have been produced on a commercial scale since 1960.

Diamonds are found in three types of deposits: alluvial gravels, glacial tills, and kimberlite pipes. The kimberlite pipes (such as those at Kimberley, South Africa) form from intrusions of magma into the Earth's crust and deliver diamonds and other rocks and minerals from the mantle. The pipes themselves are often less than 100 million years old. However, the diamonds they carry were formed 1 to 3.3 billion years ago at depths of more than about 75 miles (120 km). Diamonds found in alluvial and glacial gravels must have been released by fluvial or glacial erosion of the kimberlite matrix and then redeposited in rivers or in glacial till.

Alluvial diamond mining at Kenema, Sierra Leone *Bavaria-Verlag*.

Alluvial diamond mine Large alluvial diamond mine near the Bow River in Western Australia.

A diamond mine at Kimberley, S.Af.

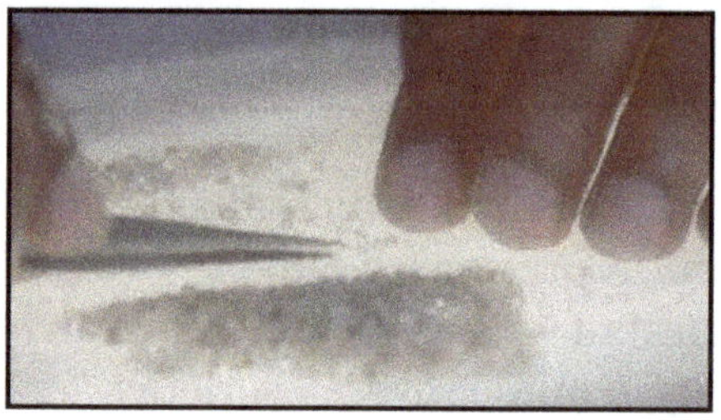

South Africa: diamond mining History of diamond mining in South Africa.

Diamonds vary from colourless to black, and they may be transparent, translucent, or opaque. Most diamonds used as gems are transparent and colourless or nearly so. Colourless or pale blue stones are most valued, but these are rare; most gem diamonds are tinged with yellow. A "fancy" diamond has a distinct body colour; red, blue, and green are rarest, and orange, violet, yellow, and yellowish green more common. Most industrial diamonds are gray or brown and are translucent or opaque, but better-quality industrial stones grade imperceptibly into poor quality gems. The colour of diamonds may be changed by exposure to intense radiation (as released in a nuclear reactor or by a particle accelerator) or by heat treatment.

A very high refractive power gives the diamond its extraordinary brilliance. A properly cut diamond will return a greater amount of light to the eye of the observer than will a gem of lesser refractive power and will thus appear more brilliant. The high dispersion gives diamonds their fire, which is caused by the separation of white light into the colours of the spectrum as it passes through the stone.

The scratch hardness of diamond is assigned the value of 10 on the Mohs scale of hardness; corundum, the mineral next to diamond in hardness, is rated as 9. Actually, diamond is very much harder than corundum; if the Mohs scale were linear, diamond's value would be about 42. The hardness of a diamond varies significantly in different directions, causing cutting and polishing of some faces to be easier than others. For detailed physical properties.

In the atomic structure of diamond, as determined by X-ray diffraction techniques, each carbon atomis linked to four equidistant neighbours throughout the crystal. This close-knit, dense, strongly bonded crystal structure yields diamond properties that differ greatly from those of graphite, native carbon's other form.

The most familiar usage of diamonds today is as gemstones used for adornment—a usage which dates back into antiquity. The dispersion of white light into spectral colors is the primary gemological characteristic of gem diamonds. In the twentieth century, gemologists have developed methods of grading diamonds and other gemstones based on the characteristics most important to their value as a gem. Four characteristics known informally as the "four Cs" are now commonly used as the basic descriptors of diamonds: carat, cut, color, and clarity. This system was developed by Gemological Institute of America in 1953 as internationally recognized standard to evaluate diamonds characteristics.

Most gem diamonds are traded on the wholesale market based on single values for each of the four Cs; for example knowing that a diamond is rated as 1.5 carats (300 mg), VS2 clarity, F color, excellent cut round brilliant, is enough to reasonably establish an expected price range. More detailed information from within each characteristic is used to determine actual market value for individual stones. Consumers who purchase individual diamonds are often advised to use the four Cs to pick the diamond that is "right" for them.

Other characteristics also influence the value and appearance of a gem diamond. These include physical characteristics such as the presence of fluorescence as well as the diamond's source and which gemological institute evaluated the diamond. *Cleanliness* also dramatically affects a diamond's beauty.

There are two major non-profit gemological associations which grade and provide reports, (informally referred to by the term *certificate* or *cert*, which is a misnomer for many grading reports) on diamonds; while carat weight and cut angles are mathematically defined, the clarity and color are judged by the trained human eye and are therefore open to slight variance in interpretation. These associations are listed below:

- Gemological Institute of America (GIA) was the first laboratory in America to issue modern diamond reports, and is held in high regard amongst gemologists for its consistent, conservative grading.
- Diamond High Council (HRD) Official certification laboratory of the Belgian diamond industry, located in Antwerp.

Within the last two decades, a number of for-profit gemological grading laboratories have also been established, many of them also based in Antwerp or New York. These entities serve to provide similar services as the non-profit associations above, but in a less expensive and more timely fashion. They produce certificates that are similar to those of the GIA.

Carat

The *carat weight* measures the mass of a diamond. One carat is defined as 200 milligrams (about 0.007 ounces avoirdupois). The *point* unit—equal to one one-hundredth of a carat (0.01 carat, or

2 mg)—is commonly used for diamonds of less than one carat. All else being equal, the price per carat increases with carat weight, since larger diamonds are both rarer and more desirable for use as gemstones.

The price per carat does not increase linearly with increasing size. Instead, there are sharp jumps around milestone carat weights, as demand is much higher for diamonds weighing just more than a milestone than for those weighing just less. As an example, a 0.99-carat diamond may have a significantly lower price per carat than a comparable 1.01-carat diamond, because of differences in demand.

A weekly diamond price list, the Rapaport Diamond Report is published by Martin Rapaport, CEO of Rapaport Group of New York, for different diamond cuts, clarity and weights. It is currently considered the de facto retail price baseline. Jewelers often trade diamonds at negotiated discounts off the Rapaport price (e.g., "R −3%").

In the wholesale trade of gem diamonds, carat is often used in denominating lots of diamonds for sale. For example, a buyer may place an order for 100 carats (20 g) of 0.5-carat (100 mg), D–F, VS2-SI1, excellent cut diamonds, indicating a wish to purchase 200 diamonds (100 carats (20 g) total mass) of those approximate characteristics. Because of this, diamond prices (particularly among wholesalers and other industry professionals) are often quoted per carat, rather than per stone.

Total carat weight (t.c.w.) is a phrase used to describe the total mass of diamonds or other gemstone in a piece of jewelry, when more than one gemstone is used. Diamond solitaire earrings, for example, are usually quoted in t.c.w. when placed for sale, indicating the mass of the diamonds in both earrings and not each individual diamond. T.c.w. is also widely used for diamond necklaces, bracelets and other similar jewelry pieces.

Clarity

Clarity is a measure of internal defects of a diamond called *inclusions*. Inclusions may be crystals of a foreign material or another diamond crystal, or structural imperfections such as tiny cracks that can appear whitish or cloudy. The number, size, color, relative location, orientation, and visibility of inclusions can all affect the relative clarity of a diamond. The Gemological Institute of America (GIA) and other organizations have developed systems to grade clarity, which are based on those inclusions which are visible to a trained professional when a diamond is viewed under 10× magnification.

Diamonds become increasingly rare when considering higher clarity gradings. Only about 20% of all diamonds mined have a clarity rating high enough for the diamond to be considered appropriate for use as a gemstone; the other 80% are relegated to industrial use. Of that top 20%, a significant portion contains one or more visible inclusions. Those that do not have a visible inclusion are known as "eye-clean" and are preferred by most buyers, although visible inclusions can sometimes be hidden under the setting in a piece of jewelry.

Most inclusions present in gem-quality diamonds do not affect the diamonds' performance or structural integrity. When set in jewelry, it may also be possible to hide certain inclusion behind mounting hardware such as prongs in a way that renders the defect invisible. However, large clouds can affect a diamond's ability to transmit and scatter light. Large cracks close to or breaking the surface may increase the likelihood of a fracture.

Diamonds are graded by the major societies on a scale ranging from flawless to imperfect.

Color

Jewelers sometimes set diamonds in groups of similar colors.

The Darya-I-Nur Diamond.

The finest quality as per color grading is totally colorless, which is graded as D color diamond across the globe, meaning it is absolutely free from any color. The next grade has a very slight trace of color, which can be observed by any expert diamond valuer/grading laboratory. However, when studded in jewellery these very light colored diamonds do not show any color or it is not possible to make out color shades. These are graded as E color or F color diamonds.

Diamonds which show very little traces of color are graded as G or H color diamonds. Slightly colored diamonds are graded as I or J or K color. A diamond can be found in any color in addition to colorless. Some of the colored diamonds, such as pink, are very rare.

A chemically pure and structurally perfect diamond is perfectly transparent with no hue, or *color*. However, in reality most gem-sized natural diamonds are imperfect. The color of a diamond may be affected by chemical impurities and structural defects in the crystal lattice. Depending on the hue and intensity of a diamond's coloration, a diamond's color can either detract from or enhance its value. For example, most white diamonds are discounted in price as a more yellow hue is detectable, while intense pink or blue diamonds (such as the Hope Diamond) can be dramatically more valuable. The Aurora Diamond Collection displays a spectacular array of naturally colored diamonds, which occur in every color of the rainbow.

Most diamonds used as gemstones are basically transparent with little tint, or *white diamonds*. The most common impurity, nitrogen, replaces a small proportion of carbon atoms in a diamond's structure and causes a yellowish to brownish tint. This effect is present in almost all white diamonds; in only the rarest diamonds is the coloration from this effect undetectable. The GIA has developed a rating system for color in white diamonds, from D to Z (with D being "colorless" and Z having a bright yellow coloration), which has been widely adopted in the industry and is universally recognized, superseding several older systems. The GIA system uses a benchmark set of natural diamonds of known color grade, along with standardized and carefully controlled lighting conditions. Diamonds with higher color grades are rarer, in higher demand, and therefore more expensive, than lower color grades. Oddly enough, diamonds graded Z are also rare, and the bright yellow color is also highly valued. Diamonds graded D–F are considered "colorless", G–J are considered "near-colorless", K–M are "slightly colored". N–Y usually appear light yellow or brown.

In contrast to yellow or brown hues, diamonds of other colors are more rare and valuable. While even a pale pink or blue hue may increase the value of a diamond, more intense coloration is usually considered more desirable and commands the highest prices. A variety of impurities and structural imperfections cause different colors in diamonds, including yellow, pink, blue, red, green, brown, and other hues. Diamonds with unusual or intense coloration are sometimes labeled "fancy" in the diamond industry. Intense yellow coloration is considered one of the fancy colors, and is separate from the color grades of white diamonds. Gemologists have developed rating systems for fancy colored diamonds, but they are not in common use because of the relative rarity of such diamonds.

Cut

Diamond cutting is the art and science of creating a gem-quality diamond out of mined rough. The *cut* of a diamond describes the manner in which a diamond has been shaped and polished from its beginning form as a rough stone to its final gem proportions. The cut of a diamond describes the quality of workmanship and the angles to which a diamond is cut. Often diamond cut is confused with "shape".

There are mathematical guidelines for the angles and length ratios at which the diamond is supposed to be cut in order to reflect the maximum amount of light. Round brilliant diamonds, the most common, are guided by these specific guidelines, though fancy cut stones are not able to be as accurately guided by mathematical specifics.

The techniques for cutting diamonds have been developed over hundreds of years, with perhaps the greatest achievements made in 1919 by mathematician and gem enthusiast Marcel Tolkowsky. He developed the round brilliant cut by calculating the ideal shape to return and scatter light when a diamond is viewed from above. The modern round brilliant has 57 facets (polished faces), counting 33 on the *crown* (the top half), and 24 on the *pavilion* (the lower half). The girdle is the thin middle part. The function of the crown is to refract light into various colors and the pavilion's function to reflect light back through the top of the diamond.

Tolkowsky's calculations included some approximations. He calculated the ideal dimensions as:

- Table percentage (corner-to-corner diameter of the table divided by overall diameter) = 53%.
- Depth percentage (overall depth divided by overall diameter) = 59.3% (not including adjustments for the culet height and girdle thickness).
- Pavilion angle (angle between the girdle and the pavilion main facets) = 40.75°.
- Crown angle (angle between the girdle and the crown's kite facets) = 34.5°.
- Pavilion depth (depth of pavilion divided by overall diameter) = 43.1%.
- Crown depth (depth of crown divided by overall diameter) = 16.2%.

The culet is the tiny point or facet at the bottom of the diamond. This should be a negligible diameter, otherwise light leaks out of the bottom. Tolkowsky's calculations included neither a culet nor a girdle. However, a girdle is required in reality in order to prevent the diamond from easily chipping in the setting. The thick part of the girdle is normally about 1.7% (of the overall diameter) thicker than the thin part of the girdle.

The further the diamond's characteristics are from the Tolkowsky's ideal, the less light will be reflected. However, there is a small range in which the diamond can be considered "ideal". Tolkowsky's calculations can be repeated for a narrow range of pavilion angles. Such calculations show a slightly larger table percentage, and a trade-off between pavilion angle and crown angle.

Today, because of the relative importance of carat weight among buyers, many diamonds are often intentionally cut poorly to increase carat weight. There is a financial premium for a diamond that weighs the desirable 1.0 carat (200 mg), so often the girdle is made thicker or the depth is increased. Neither of these changes makes the diamond *appear* any larger, and both greatly reduce the sparkle of the diamond. (A poorly cut 1.0-carat (200 mg) diamond may have the same diameter and appear as large as a 0.85-carat (170 mg) diamond.) The *depth percentage* is the overall quickest indication of the quality of the cut of a round brilliant. "Ideal" round brilliant diamonds should not have a depth percentage greater than 62.5%. Another quick indication is the *overall diameter*. Typically a round brilliant 1.0-carat (200 mg) diamond should have a diameter of about 6.5 mm (0.26 in). Mathematically, the diameter in millimeters of a round brilliant should approximately equal to 6.5 times the cube root of carat weight, or 11.1 times the cube root of gram weight, or 1.4 times the cube root of point weight.

Shape

Diamonds do not show all of their beauty as rough stones; instead, they must be cut and polished to exhibit the characteristic fire and brilliance that diamond gemstones are known for. Diamonds are cut into a variety of shapes that are generally designed to accentuate these features.

Diamonds which are not cut into a round brilliant shape are known as "fancy cuts." Popular fancy cuts include the *baguette* (French, meaning *rod* or loaf of bread), *marquise*, *princess cut* (square outline), *heart*, *briolette* (a form of the rose cut), and *pear* cuts. Newer cuts that have been introduced into the jewelry industry are the "cushion" "radiant" (similar to princess cuts, but with rounded edges instead of square edges) and Asscher cuts. Many fancy colored diamonds are now being cut according to these new styles. Generally speaking, these "fancy cuts" are not held to the same strict standards as Tolkowsky-derived round brilliants and there are less specific mathematical guidelines of angles which determine a well-cut stone. Cuts are influenced heavily by fashion: the baguette cut—which accentuates a diamond's luster and downplays its fire—was popular during the Art Deco period, whereas the princess cut — which accentuates a diamond's fire rather than its luster — is currently gaining popularity. The princess cut is also popular amongst diamond cutters: of all the cuts, it wastes the least of the original crystal. The past decades have seen the development of new diamond cuts, often based on a modification of an existing cut. Some of these include extra facets. These newly developed cuts are viewed by many as more of an attempt at brand differentiation by diamond sellers, than actual improvements to the state of the art.

Quality

The quality of a diamond's cut is widely considered the most important of the four Cs in determining the beauty of a diamond; indeed, it is commonly acknowledged that a well-cut diamond can appear to be of greater carat weight, and have clarity and color appear to be of better grade than they actually are. The skill with which a diamond is cut determines its ability to reflect and refract light.

In addition to carrying the most importance to a diamond's quality as a gemstone, the cut is also the most difficult to quantitatively judge. A number of factors, including proportion, polish, symmetry, and the relative angles of various facets, are determined by the quality of the cut and can affect the performance of a diamond. A diamond with facets cut only a few degrees out of alignment can result in a poorly performing stone. For a round brilliant cut, there is a balance between "brilliance" and "fire". When a diamond is cut for too much "fire", it looks like a cubic zirconia, which gives off much more "fire" than real diamond. A well-executed round brilliant cut should reflect light upwards and make the diamond appear white when viewed from the top. An inferior cut will produce a stone that appears dark at the center and in extreme cases the setting may be seen through the top of the diamond as shadows.

Several different theories on the "ideal" proportions of a diamond have been and continue to be advocated by various owners of patents on machines to view how well a diamond is cut. These advocate a shift away from grading cut by the use of various angles and proportions toward measuring the performance of a cut stone. A number of specially modified viewers and machines have been developed toward this end. Hearts and Arrows viewers test for the "hearts and arrows" characteristic pattern observable in stones exhibiting high symmetry and particular cut angles. Closely related to Hearts and Arrows viewers is the ASET which tests for light leakage, light return, and proportions. The ASET (and computer simulations of the ASET) are used to test for AGS cut grade. Proponents of these machines argue they help sellers demonstrate the light performance of the diamond in addition to the traditional 4 Cs. Detractors, however, see these machines as marketing tools rather than scientific ones. The GIA has developed a set of criteria for grading the cut of round brilliant stones that is now the standard in the diamond industry and is called Facetware.

Process

The 253-carat (50.6 g) Oppenheimer Diamond, at a 2001 diamond exhibition in Paris. An uncut diamond does not show its prized optical properties.

The process of shaping a rough diamond into a polished gemstone is both an art and a science. The choice of cut is often decided by the original shape of the rough stone, location of the inclusions and flaws to be eliminated, the preservation of the weight, popularity of certain shapes amongst consumers and many other considerations. The round brilliant cut is preferred when the crystal is an octahedron, as often two stones may be cut from one such crystal. Oddly shaped crystals such as macles are more likely to be cut in a *fancy cut*—that is, a cut other than the round brilliant—which the particular crystal shape lends itself to.

Even with modern techniques, the cutting and polishing of a diamond crystal always results in a dramatic loss of weight; rarely is it less than 50%. Sometimes the cutters compromise and accept lesser proportions and symmetry in order to avoid inclusions or to preserve the carat rating. Since the per carat price of diamond shifts around key milestones (such as 1.00 carat (200 mg)), many one-carat diamonds are the result of compromising cut for carat. Some jewelry experts advise consumers to buy a 0.99-carat (198 mg) diamond for its better price or buy a 1.10-carat (220 mg) diamond for its better cut, avoiding a 1.00-carat (200 mg) diamond which is more likely to be a poorly cut stone.

Light Performance

In the gem trade, the term light performance is used to describe how well a polished diamond will return light to the viewer. There are three light properties which are described in relation to light performance: brilliance, fire, and scintillation. Brilliance refers to the white light reflections from the external and internal facet surfaces. Fire refers to the spectral colors which are produced as a result of the diamond dispersing the white light. Scintillation refers to the small flashes of light that are seen when the diamond, light source or the viewer is moved. A diamond that is cut and polished to produce a high level of these qualities is said to be high in light performance.

The setting diamonds are placed in also affect the performance of light through a diamond. The three most commonly used settings are: Prong, Bezel, and Channel. Prong settings are the most popular setting for diamond jewelry. The prong setting consists of four or six 'claws' that cradle the diamond, allowing the maximum amount of light to enter from all angles, allowing the diamonds to appear larger and more brilliant. In bezel settings the diamond or gemstone is completely surrounded by a rim of metal, which can be molded into any shape to accommodate the stone. Used to set earrings, necklaces, bracelets, and rings, bezel settings can have open or closed backs, and generally can be molded to allow a lot of light to pass through. Channel settings set the stones right next to each other with no metal separating them. This setting is mostly used in wedding and anniversary bands. The outer ridge is then worked over the edges of the stones to create a smooth exterior surface. This also protects the girdle area of the stone.

Fluorescence

About a third of all diamonds will glow under ultraviolet light, usually a blue color which may be noticeable under a black light or strong sunlight. According to the GIA, who reviewed a random sample of 26,010 natural diamonds, 65% of the diamonds in the sample had no fluorescence. Of the 35% that did have fluorescence, 97% had blue fluorescence of which 38% had faint blue fluorescence and 62% had fluorescence that ranged from medium to very strong blue. Other colors diamonds can fluoresce are green, yellow, and red but are very rare and are sometimes a combination of the colors such as blue-green or orange. Some diamonds with "very strong" fluorescence can have a "milky" or "oily" look to them, but they are also very rare and are termed "overblues." Their study concluded that with the exception of "overblues" and yellow fluorescent diamonds, fluorescence had little effect on transparency and that the strong and very strong blue fluorescent diamonds on average had better color appearance than non-fluorescent stones. Since blue is a complementary color to yellow and can appear to cancel it out, strong blue fluorescence had especially better color appearance with lower color graded diamonds that have a slight yellowish tint such as I or J color but had little effect on the more colorless D, E and F color grades.

Cleaning

Cleanliness significantly affects a diamond's beauty. A clean diamond is more brilliant and fiery than the same diamond when it is "dirty". Dirt or grease on the top of a diamond reduces its luster. Water, dirt, or grease on the bottom of a diamond interferes with the diamond's brilliance and fire. Even a thin film absorbs some light that could have been reflected to the viewer. Colored dye or smudges can affect the perceived color of a diamond. Historically, some jewelers' stones were misgraded because of smudges on the girdle, or dye on the culet. Current practice is to clean a diamond thoroughly before grading its color.

Maintaining a clean diamond can sometimes be difficult as jewelry settings can obstruct cleaning, and oils, grease, and other hydrophobic materials adhere well to a diamond. Many jewelers use steam cleaners. Some jewelers provide their customers with ammonia-based cleaning kits; ultrasonic cleaners are also popular.

Sapphire

Sapphire is the second hardest crystal that Mother Earth has given us. Second only to the Diamond. Many of the large Sapphire deposits are literally the remains of metamorphic or igneous veins of granite and marble weathered away into stream beds and mountain valleys. Sapphire being stronger than granite or marble survive the weathering and large amounts of the gemstone are simply gathered from gem gravels in Burma, Thailand, Australia, North America, Brazil, Kashmir and Africa.

Sapphire's beauty and enduring strength historically made it quite popular as a talisman among royalty. Legend tells us that medieval kings wore the stone not only to protect themselves from harm, but also to prevent jealousy among their court. A Sapphire worn to the signing of a treaty was believed to ensure a true reconciliation and prevent further bloodshed.

Natural Sapphires

An uncut, rough yellow sapphire found at the Spokane Sapphire Mine near Helena, Montana.

Sapphire is one of the two gem-varieties of corundum, the other being ruby(defined as corundum in a shade of red). Although blue is the best-known sapphire color, they occur in other colors, including gray and black, and they can be colorless. A pinkish orange variety of sapphire is called padparadscha.

Significant sapphire deposits are found in Eastern Australia, Thailand, Sri Lanka, China (Shandong), Madagascar, East Africa, and in North Americain a few locations, mostly in Montana. Sapphire and rubies are often found in the same geological setting.

Every sapphire mine produces a wide range of quality – and origin is not a guarantee of quality. For sapphire, Kashmirreceives the highest premium although Burma, Sri Lanka, and Madagascar also produce large quantities of fine quality gems.

The cost of natural sapphires varies depending on their color, clarity, size, cut, and overall quality. For gems of exceptional quality, an independent determination from a respected laboratory such as the GIA, AGL or Gubelin of origin often adds to value.

Blue Sapphire

Teardrop-shaped blue sapphire.

Gemstone color can be described in terms of hue, saturation, and tone. Hue is commonly understood as the "color" of the gemstone. Saturation refers to the vividness or brightness of the hue, and tone is the lightness to darkness of the hue. Blue sapphire exists in various mixtures of its primary (blue) and secondary hues, various tonal levels (shades) and at various levels of saturation (vividness).

Blue sapphires are evaluated based upon the purity of their primary hue. Purple, violet, and green are the most common secondary hues found in blue sapphires. Violet and purple can contribute to the overall beauty of the color, while green is considered to be distinctly negative. Blue sapphires with up to 15% violet or purple are generally said to be of fine quality. Gray is the normal saturation modifier or mask found in blue sapphires. Gray reduces the saturation or brightness of the hue, and therefore has a distinctly negative effect.

The color of fine blue sapphires may be described as a vivid medium dark violet to purplish blue where the primary blue hue is at least 85% and the secondary hue no more than 15%, without the least admixture of a green secondary hue or a gray mask.

The 423-carat (84.6 g) Logan sapphire in the National Museum of Natural History, in Washington, D.C., is one of the largest faceted gem-quality blue sapphires in existence.

Dark blue sapphire, probably of Australian origin, showing the brilliant surface luster typical of faceted corundum gemstones.

Sapphires of other Colors

Sapphires in colors other than blue are called "fancy" or "parti colored" sapphires.

Fancy sapphires are often found in yellow, orange, green, brown, purple and violet hues.

Particolored sapphires are those stones which exhibit two or more colors within a single stone. Australia is the largest source of particolored sapphires; they are not commonly used in mainstream jewelry and remain relatively unknown. Particolored sapphires cannot be created synthetically and only occur naturally.

Colorless sapphires have historically been used as diamond substitutes in jewelry.

Pink Sapphires

Pink sapphires occur in shades from light to dark pink, and deepen in color as the quantity of chromium increases. The deeper the pink color, the higher their monetary value. In the United States, a minimum color saturation must be met to be called a ruby, otherwise the stone is referred to as a *pink sapphire.*

Pink sapphire (American terminology, may be classified as rubyoutside of the United States).

Padparadscha

Padparadscha is a delicate, light to medium toned, pink-orange to orange-pink hued corundum, originally found in Sri Lanka, but also found in deposits in Vietnam and parts of East Africa. Padparadscha sapphires are rare; the rarest of all is the totally natural variety, with no sign of artificial treatment.

Faceted padparadscha.

The name is derived from the Sanskrit "padma ranga" (padma = lotus; ranga = color), a color akin to the lotus flower (*Nelumbo nucifera*).

Natural padparadscha sapphires often draw higher prices than many of even the finest blue sapphires. Recently, more sapphires of this color have appeared on the market as a result of a new artificial treatment method called "lattice diffusion".

Star Sapphire

A star sapphire is a type of sapphire that exhibits a star-like phenomenon known as asterism; red stones are known as "star rubies". Star sapphires contain intersecting needle-like inclusions following the underlying crystal structure that causes the appearance of a six-rayed "star"-shaped pattern when viewed with a single overhead light source. The inclusion is often the mineral rutile, a mineral composed primarily of titanium dioxide. The stones are cut *en cabochon*, typically with the center of the star near the top of the dome. Occasionally, twelve-rayed stars are found, typically because two different sets of inclusions are found within the same stone, such as a combination of fine needles of rutile with small platelets of hematite; the first results in a whitish star and the second results in a golden-colored star. During crystallisation, the two types of inclusions become preferentially oriented in different directions within the crystal, thereby forming two six-rayed stars that are superimposed upon each other to form a twelve-rayed star. Misshapen stars or 12-rayed stars may also form as a result of twinning. The inclusions can alternatively produce a "cat's eye" effect if the ‹face-up' direction of the cabochon's dome is oriented perpendicular to the crystal's c-axis rather than parallel to it. If the dome is oriented in between these two directions, an 'off-center' star will be visible, offset away from the high point of the dome.

Star sapphire.

The Star of Adam is the largest blue star sapphire which weighs 1404.49 carats. The gem was mined in the city of Ratnapura, southern Sri Lanka. The Black Star of Queensland, the second largest gem-quality star sapphire in the world, weighs 733 carats. The Star of India mined in Sri Lanka and weighing 563.4 carats is thought to be the third-largest star sapphire, and is currently on display at the American Museum of Natural History in New York City. The 182-carat Star of Bombay, mined in Sri Lanka and located in the National Museum of Natural History in Washington, D.C., is another example of a large blue star sapphire. The value of a star sapphire depends not only on the weight of the stone, but also the body color, visibility, and intensity of the asterism.

Color Change Sapphire

A rare variety of natural sapphire, known as color-change sapphire, exhibits different color in different light. Color change sapphires are blue in outdoor light and purple under incandescent indoor light, or green to gray-green in daylight and pink to reddish-violet in incandescent light. Color change sapphires come from a variety of locations, including Thailand and Tanzania. The color-change effect is caused by the interaction of the sapphire, which absorbs specific wavelengths of light, and the light-source, whose spectral output varies depending upon the illuminant. Transition-metal impurities in the sapphire, such as chromium and vanadium, are responsible for the color change.

Certain synthetic color-change sapphires have a similar color change to the natural gemstone alexandrite and they are sometimes marketed as "alexandrium" or "synthetic alexandrite". However, the latter term is a misnomer: synthetic color-change sapphires are, technically, not synthetic alexandrites but rather alexandrite *simulants*. This is because genuine alexandrite is a variety of chrysoberyl: not sapphire, but an entirely different mineral.

Source of Color

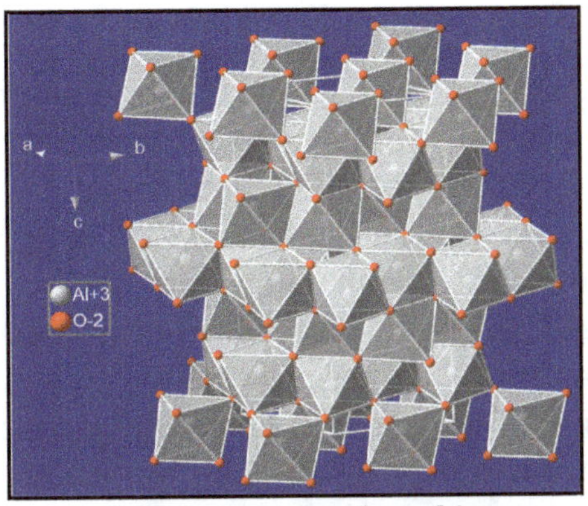

Crystal structure of sapphire.

Sapphire ring made circa 1940.

Rubies are corundum which contain chromium impurities that absorb yellow-green light and result in deeper ruby redcolor with increasing content. Purple sapphires contain trace amounts of vanadium and come in a variety of shades. Corundum that contains ~0.01% of titanium is colorless. If trace amounts of iron are present, a very pale yellow to green color may be seen. However, if both titanium and iron impurities are present together, and in the correct valence states, the result is a deep-blue color.

Unlike localized ("intra-atomic") absorption of light which causes color for chromium and vanadium impurities, blue color in sapphires comes from intervalence charge transfer, which is the transfer of an electron from one transition-metal ion to another via the conduction or valence band. The iron can take the form Fe^{2+} or Fe^{3+}, while titanium generally takes the form Ti^{4+}. If Fe^{2+} and Ti^{4+} ions are substituted for Al^{3+}, localized areas of charge imbalance are created. An electron transfer from Fe^{2+} and Ti^{4+} can cause a change in the valence state of both. Because of the valence change there is a specific change in energy for the electron, and electromagnetic energy is absorbed. The wavelength of the energy absorbed corresponds to yellow light. When this light is subtracted from incident white light, the complementary color blue results. Sometimes when atomic spacing is different in different directions there is resulting blue-green dichroism.

Intervalence charge transfer is a process that produces a strong colored appearance at a low percentage of impurity. While at least 1% chromium must be present in corundum before the deep red ruby color is seen, sapphire blue is apparent with the presence of only 0.01% of titanium and iron.

Treatments

Sapphires can be treated by several methods to enhance and improve their clarity and color. It is common practice to heat natural sapphires to improve or enhance color. This is done by heating the sapphires in furnaces to temperatures between 500 and 1,800°C (932 and 3,272 °F) for several hours, or by heating in a nitrogen-deficient atmosphere oven for seven days or more. Upon heating, the stone becomes more blue in color, but loses some of the rutile inclusions (silk). When high temperatures are used, the stone loses all silk (inclusions) and it becomes clear under magnification. The inclusions in natural stones are easily seen with a jeweler's loupe. Evidence of sapphire and other gemstones being subjected to heating goes back at least to Roman times. Unheated natural stones are somewhat rare and will often be sold accompanied by a certificate from an independent gemological laboratory attesting to "no evidence of heat treatment".

Yogo sapphire.

Yogo sapphires sometimes do not need heat treating because their cornflower blue coloring is uniform and deep, they are generally free of the characteristic inclusions, and they have high uniform clarity. When Intergem Limited began marketing the Yogo in the 1980s as the world's only guaranteed untreated sapphire, heat treatment was not commonly disclosed; by 1982 the heat treatment became a major issue. At that time, 95% of all the world's sapphires were being heated to enhance their natural color. Intergem's marketing of guaranteed untreated Yogos set them against many in the gem industry.

Diffusion treatments are used to add impurities to the sapphire to enhance color. Typically beryllium is diffused into a sapphire under very high heat, just below the melting point of the sapphire. Initially orange sapphires were created, although now the process has been advanced and many colors of sapphire are often treated with beryllium. The colored layer can be removed when stones chip or are repolished or refaceted, depending on the depth of the impurity layer. Treated padparadschas may be very difficult to detect, and many stones are certified by gemological labs (*e.g.*, Gubelin, SSEF, AGTA).

According to United States Federal Trade Commission guidelines, disclosure is required of any mode of enhancement that has a significant effect on the gem's value.

There are several ways of treating sapphire. Heat-treatment in a reducing or oxidising atmosphere (but without the use of any other added impurities) is commonly used to improve the color of sapphires, and this process is sometimes known as "heating only" in the gem trade. In contrast, however, heat treatment combined with the deliberate addition of certain specific impurities (e.g. beryllium, titanium, iron, chromium or nickel, which are absorbed into the crystal structure of the sapphire) is also commonly performed, and this process can be known as "diffusion" in the gem trade. However, despite what the terms "heating only" and "diffusion" might suggest, both of these categories of treatment actually involve diffusion processes.

Mining

Sapphire from Madagascar.

Sapphires are mined from alluvial deposits or from primary underground workings. Commercial mining locations for sapphire and ruby include (but are not limited to) the following countries: Afghanistan, Australia, Myanmar/Burma, Cambodia, China, Colombia, India, Kenya, Laos, Madagascar, Malawi, Nepal, Nigeria, Pakistan, Sri Lanka, Tajikistan, Tanzania, Thailand, United States, and Vietnam. Sapphires from different geographic locations may have different appearances or chemical-impurity concentrations, and tend to contain different types of microscopic inclusions. Because of this, sapphires can be divided into three broad categories: classic metamorphic, non-classic metamorphic or magmatic, and classic magmatic.

Sapphires from certain locations, or of certain categories, may be more commercially appealing than others,particularly classic metamorphic sapphires from Kashmir, Burma, or Sri Lanka that have not been subjected to heat-treatment.

The Logan sapphire, the Star of India, and the Star of Bombay originate from Sri Lankan mines. Madagascar is the world leader in sapphire production (as of 2007) specifically its deposits in and around the town of Ilakaka. Prior to the opening of the Ilakaka mines, Australia was the largest producer of sapphires (such as in 1987). In 1991 a new source of sapphires was discovered in Andranondambo, southern Madagascar. That area has been exploited for its sapphires started in 1993, but it was practically abandoned just a few years later—because of the difficulties in recovering sapphires in their bedrock.

In North America, sapphires have been mined mostly from deposits in Montana: fancies along the Missouri River near Helena, Montana, Dry Cottonwood Creek near Deer Lodge, Montana, and Rock Creek near Philipsburg, Montana. Fine blue Yogo sapphires are found at Yogo Gulch west of Lewistown, Montana. A few gem-grade sapphires and rubies have also been found in the area of Franklin, North Carolina.

The sapphire deposits of Kashmir are well known in the gem industry, although their peak production took place in a relatively short period at the end of the nineteenth and early twentieth centuries. They have a superior cornflower blue hue to them with a mysterious and almost sleepy quality, described by some gem enthusiasts as 'blue velvet". Kashmir-origin contributes meaningfully to the value of a sapphire, and most corundum of Kashmir origin can be readily identified by its characteristic silky appearance and exceptional hue. The unique blue appears lustrous under any kind of light, unlike non-Kashmir sapphires which may appear purplish or grayish in comparison. Sotheby's has been in the forefront overseeing record-breaking sales of Kashmir sapphires worldwide. In October 2014, Sotheby's Hong Kong achieved consecutive per-carat price records for Kashmir sapphires - first with the 12.00 carat Cartier sapphire ring at US$193,975 per carat, then with a 17.16 carat sapphire at US$236,404, and again in June 2015 when the per-carat auction record was set at US$240,205. At present, the world record price-per-carat for sapphire at auction is held by a sapphire from Kashmir in a ring, which sold in October 2015 for approximately US$242,000 per carat (HK$52,280,000 in total, including buyer's premium, or more than US$6.74 million).

Synthetic Sapphire

Synthetic sapphire.

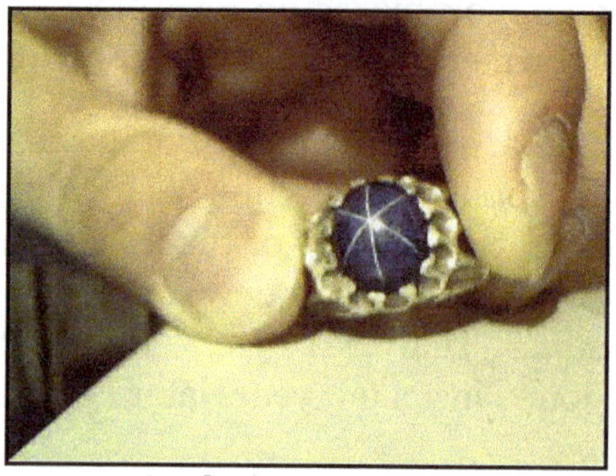

Synthetic star sapphire.

In 1902, the French chemist Auguste Verneuil developed a process for producing synthetic sapphire crystals. In the Verneuil process, named after him, fine alumina powder is added to an oxyhydrogen flame, and this is directed downward against a mantle. The alumina in the flame is slowly deposited, creating a teardrop shaped "boule" of sapphire material. Chemical dopants can be added to create artificial versions of the ruby, and all the other natural colors of sapphire, and in addition, other colors never seen in geological samples. Artificial sapphire material is identical to natural sapphire, except it can be made without the flaws that are found in natural stones. The disadvantage of Verneuil process is that the grown crystals have high internal strains. Many methods of manufacturing sapphire today are variations of the Czochralski process, which was invented in 1916 by Polish chemist Jan Czochralski. In this process, a tiny sapphire seed crystal is dipped into a crucible made of the precious metal iridium or molybdenum, containing molten alumina, and then slowly withdrawn upward at a rate of 1 to 100 mm per hour. The alumina crystallizes on the end, creating long carrot-shaped boules of large size up to 200 kg in mass.

Synthetic sapphire is also produced industrially from agglomerated aluminium oxide, sintered and fused (such as by hot isostatic pressing) in an inert atmosphere, yielding a transparent but slightly porous polycrystalline product.

In 2003, the world's production of synthetic sapphire was 250 tons (1.25×10^9 carats), mostly by the United States and Russia. The availability of cheap synthetic sapphire unlocked many industrial uses for this unique material.

The first laser was made with a rod of synthetic ruby. Titanium-sapphire lasers are popular due to their relatively rare capacity to be tuned to various wavelengths in the red and near-infrared region of the electromagnetic spectrum. They can also be easily mode-locked. In these lasers a synthetically produced sapphire crystal with chromium or titaniumimpurities is irradiated with intense light from a special lamp, or another laser, to create stimulated emission.

Common Applications

Windows

Cermax xenon arc lamp with synthetic sapphire output window.

Synthetic sapphire—sometimes referred to as *sapphire glass*—is commonly used as a window material, because it is both highly transparent to wavelengths of light between 150 nm (UV) and 5500 nm (IR) (the visible spectrum extends about 380 nm to 750 nm), and extraordinarily scratch-resistant.

The key benefits of sapphire windows are:

- Very wide optical transmission band from UV to near-infrared, (0.15–5.5 μm).

- Significantly stronger than other optical materials or standard glass windows.
- Highly resistant to scratching and abrasion (9 on the Mohs scale of mineral hardness scale, the 3rd hardest natural substance next to moissanite and diamonds).
- Extremely high melting temperature (2030° C).

Single crystal sapphire boule grown by the Kyropoulos method.
Approximately 200 mm diameter and approximately 30 kg.

Some sapphire-glass windows are made from pure sapphire boules that have been grown in a specific crystal orientation, typically along the optical axis, the c-axis, for minimum birefringence for the application.

The boules are sliced up into the desired window thickness and finally polished to the desired surface finish. Sapphire optical windows can be polished to a wide range of surface finishes due to its crystal structure and its hardness. The surface finishes of optical windows are normally called out by the scratch-dig specifications in accordance with the globally adopted MIL-O-13830 specification.

The sapphire windows are used in both high pressure and vacuum chambers for spectroscopy, crystals in various watches, and windows in grocery store barcode scanners since the material's exceptional hardness and toughness makes it very resistant to scratching.

It is used for end windows on some high-powered laser tubes as its wide-band transparency and thermal conductivity allow it to handle very high power densities in the infra-red or UV spectrum without degrading due to heating.

Along with zirconia and aluminium oxynitride, synthetic sapphire is used for shatter resistant windows in armored vehicles and various military body armor suits, in association with composites.

One type of xenon arc lamp – originally called the "Cermax" and now known generically as the "ceramic body xenon lamp" – uses sapphire crystal output windows. This product tolerates higher thermal loads and thus higher output powers when compared with conventional Xe lamps with pure silica window.

Use as Substrate for Semiconducting Circuits

Thin sapphire wafers were the first successful use of an insulating substrate upon which to deposit silicon to make the integrated circuits known as silicon on sapphire or "SOS"; now other substrates can also be used for the class of circuits known more generally as silicon on insulator. Besides its excellent electrical insulating properties, sapphire has high thermal conductivity. CMOS chips on sapphire are especially useful for high-power radio-frequency (RF) applications such as those found in cellular telephones, public-safety band radios, and satellite communication systems. "SOS" also allows for the monolithic integration of both digital and analog circuitry all on one IC chip, and the construction of extremely low power circuits.

In one process, after single crystal sapphire boules are grown, they are core-drilled into cylindrical rods, and wafers are then sliced from these cores.

Wafers of single-crystal sapphire are also used in the semiconductor industry as substrates for the growth of devices based on gallium nitride (GaN). The use of sapphire significantly reduces the cost, because it has about one-seventh the cost of germanium. Gallium nitride on sapphire is commonly used in blue light-emitting diodes (LEDs).

Use for Endoprostheses

Monocrystalline sapphire is fairly biocompatible and the exceptionally low wear of sapphire–metal pairs has led to the introduction (in Ukraine) of sapphire monocrystals for hip joint endoprostheses.

Ruby

A ruby is a pink to blood-red colored gemstone, a variety of the mineral corundum (aluminium oxide). Other varieties of gem-quality corundum are called sapphires. Ruby is one of the traditional cardinal gems, together with amethyst, sapphire, emerald, and diamond. The word *ruby* comes from *ruber*, Latin for red. The color of a ruby is due to the element chromium.

The quality of a ruby is determined by its color, cut, and clarity, which, along with carat weight, affect its value. The brightest and most valuable shade of red called blood-red or pigeon blood, commands a large premium over other rubies of similar quality. After color follows clarity: similar to diamonds, a clear stone will command a premium, but a ruby without any needle-like rutile inclusions may indicate that the stone has been treated. Ruby is the traditional birthstone for July and is usually more pink than garnet, although some rhodolite garnets have a similar pinkish hue to most rubies. The world's most valuable ruby is the Sunrise Ruby.

Physical Properties

Rubies have a hardness of 9.0 on the Mohs scale of mineral hardness. Among the natural gems only moissanite and diamond are harder, with diamond having a Mohs hardness of 10.0 and moissanite falling somewhere in between corundum (ruby) and diamond in hardness. Sapphire, ruby, and pure corundum are α-alumina, the most stable form of Al_2O_3, in which 3 electrons leave each aluminum ion to join the regular octahedral group of six nearby O^{2-} ions; in pure corundum this leaves all of the aluminum ions with a very stable configuration of no unpaired electrons or unfilled energy levels, and the crystal is perfectly colorless.

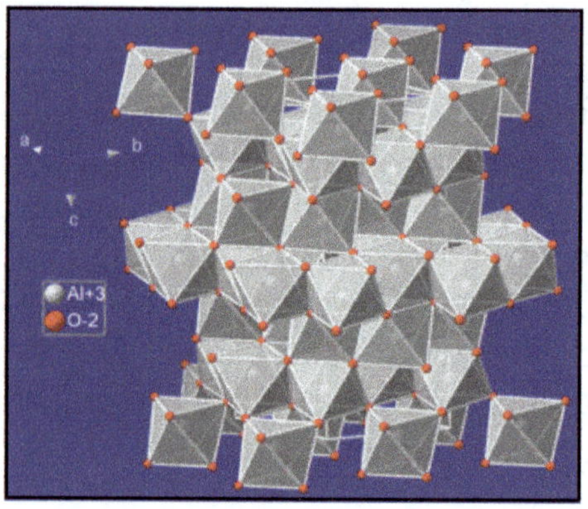

Crystal structure of rubies.

When a chromium atom replaces an occasional aluminum atom, it too loses 3 electrons to become a chromium^{3+} ion to maintain the charge balance of the Al_2O_3 crystal. However, the Cr^{3+} ions are larger and have electron orbitals in different directions than aluminum. The octahedral arrangement of the O^{2-} ions is distorted, and the energy levels of the different orbitals of those Cr^{3+} ions are slightly altered because of the directions to the O^{2-} ions. Those energy differences correspond to absorption in the ultraviolet, violet, and yellow-green regions of the spectrum.

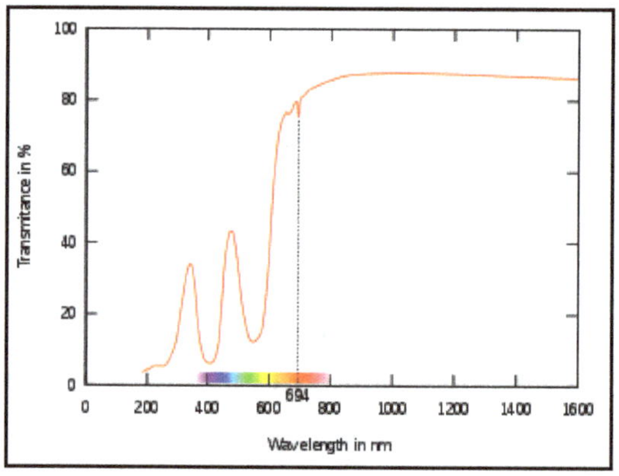

Transmittance of ruby in optical and near-IR spectra. Note the two broad violet and yellow-green absorption bands and one narrow absorption band at the wavelength of 694 nm, which is the wavelength of the ruby laser.

If one percent of the aluminum ions are replaced by chromium in ruby, the yellow-green absorption results in a red color for the gem. Additionally, absorption at any of the above wavelengths stimulates fluorescent emission of 694-nanometer-wavelength red light, which adds to its red color and perceived luster.

After absorbing short-wavelength light, there is a short interval of time when the crystal lattice of ruby is in an excited state before fluorescence occurs. If 694-nanometer photons pass through the crystal during that time, they can stimulate more fluorescent photons to be emitted in-phase with them, thus strengthening the intensity of that red light. By arranging mirrors or other means to pass emitted light repeatedly through the crystal, a ruby laser in this way produces a very high intensity of coherent red light.

All natural rubies have imperfections in them, including color impurities and inclusions of rutile needles known as "silk". Gemologists use these needle inclusions found in natural rubies to distinguish them from synthetics, simulants, or substitutes. Usually, the rough stone is heated before cutting. These days, almost all rubies are treated in some form, with heat treatment being the most common practice. Untreated rubies of high quality command a large premium.

Some rubies show a three-point or six-point asterism or "star". These rubies are cut into cabochons to display the effect properly. Asterisms are best visible with a single-light source and move across the stone as the light moves or the stone is rotated. Such effects occur when light is reflected off the "silk" (the structurally oriented rutile needle inclusions) in a certain way. This is one example where inclusions increase the value of a gemstone. Furthermore, rubies can show color changes—though this occurs very rarely—as well as chatoyancy or the "cat's eye" effect.

Ruby vs. Pink Sapphire

Generally, gemstone-quality corundum in all shades of red, including pink, are called rubies. However, in the United States, a minimum color saturation must be met to be called a ruby; otherwise, the stone will be called a pink sapphire.Drawing a distinction between rubies and pink sapphires is relatively new, having arisen sometime in the 20th century. Often, the distinction between ruby and pink sapphire is not clear and can be debated. As a result of the difficulty and subjectiveness of such distinctions, trade organizations such as the International Colored Gemstone Association (ICGA) have adopted the broader definition for ruby which encompasses its lighter shades, including pink.

Natural Occurrence

The Mogok Valley in Upper Myanmar (Burma) was for centuries the world's main source for rubies. That region has produced some exceptional rubies, however in recent years few good rubies have been found. In central Myanmar, the area of Mong Hsu began producing rubies during the 1990s and rapidly became the world's main ruby mining area. The most recently found ruby deposit in Myanmar is in Namya (Namyazeik) located in the northern state of Kachin.

Historically, rubies have also been mined in Thailand, in the Pailin and Samlout District of Cambodia, as well as in Afghanistan, Australia, Brazil, Colombia, India, Namibia, Japan, and Scotland; after the Second World War ruby deposits were found in Madagascar, Nepal, Pakistan, Tajikistan, Tanzania, and Vietnam. In Sri Lanka, lighter shades of rubies (often "pink sapphires") are more commonly found. The Republic of Macedonia is the only country in mainland Europe to have naturally occurring rubies. They can mainly be found around the city of Prilep. Macedonian rubies have a unique raspberry color. The ruby is also included on the Macedonian coat of arms. A few rubies have been found in the U.S. states of Montana, North Carolina, South Carolina and Wyoming.

Spinel, another red gemstone, is sometimes found along with rubies in the same gem gravel or marble. Red spinels may be mistaken for rubies by those lacking experience with gems. However, the finest red spinels can have values approaching that of an average ruby.

Factors Affecting Value

Rubies, as with other gemstones, are graded using criteria known as the four Cs, namely color, cut, clarity and carat weight. Rubies are also evaluated on the basis of their geographic origin.

Color: In the evaluation of colored gemstones, color is the most important factor. Color divides into three components: *hue, saturation* and *tone*. Hue refers to color as we normally use the term. Transparent gemstones occur in the *pure spectral hues* of red, orange, yellow, green, blue, violet. In nature, there are rarely pure hues, so when speaking of the hue of a gemstone, we speak of primary and secondary and sometimes tertiary hues. Ruby is defined to be red. All other hues of the gem species corundum are called sapphire. Ruby may exhibit a range of secondary hues, including orange, purple, violet, and pink.

Natural ruby with inclusions

A cut pink ruby

Purple rubies

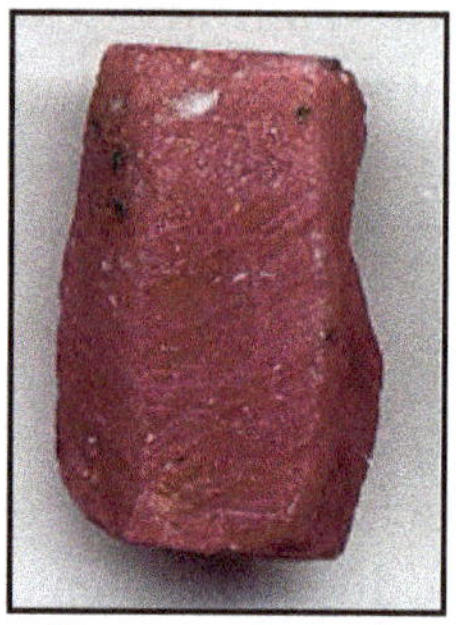
A naturally occurring ruby crystal

Victorian ruby and diamond brooch

The finest ruby is described as being a vivid medium-dark toned red. Secondary hues add an additional complication. Pink, orange, and purple are the normal secondary hues in ruby. Of the three, purple is preferred because it reinforces the red, making it appear richer. Purple also occupies a position on the color wheel halfway between red and blue. When a purplish-red ruby is set in yellow, the yellow neutralizes its complement blue, leaving the stone appearing to be pure red in the setting.

Treatments and Enhancements

Improving the quality of gemstones by treating them is common practice. Some treatments are used in almost all cases and are therefore considered acceptable. During the late 1990s, a large supply of low-cost materials caused a sudden surge in supply of heat-treated rubies, leading to a downward pressure on ruby prices.

Improvements used include color alteration, improving transparency by dissolving rutile inclusions, healing of fractures (cracks) or even completely filling them.

The most common treatment is the application of heat. Most rubies at the lower end of the market are heat treated to improve color, remove *purple tinge*, blue patches, and silk. These heat treatments typically occur around temperatures of 1800°C (3300 °F). Some rubies undergo a process of low tube heat, when the stone is heated over charcoal of a temperature of about 1300°C (2400 °F) for 20 to 30 minutes. The silk is partially broken, and the color is improved.

Another treatment, which has become more frequent in recent years, is lead glass filling. Filling the fractures inside the ruby with lead glass (or a similar material) dramatically improves the transparency of the stone, making previously unsuitable rubies fit for applications in jewelry. The process is done in four steps:

1. The rough stones are pre-polished to eradicate all surface impurities that may affect the process.
2. The rough is cleaned with hydrogen fluoride.
3. The first heating process during which no fillers are added. The heating process eradicates impurities inside the fractures. Although this can be done at temperatures up to 1400 °C (2500 °F) it most likely occurs at a temperature of around 900 °C (1600 °F) since the rutile silk is still intact.
4. The second heating process in an electrical oven with different chemical additives. Different solutions and mixes have shown to be successful, however mostly lead-containing glass-powder is used at present. The ruby is dipped into oils, then covered with powder, embedded on a tile and placed in the oven where it is heated at around 900 °C (1600 °F) for one hour in an oxidizing atmosphere. The orange colored powder transforms upon heating into a transparent to yellow-colored paste, which fills all fractures. After cooling the color of the paste is fully transparent and dramatically improves the overall transparency of the ruby.

If a color needs to be added, the glass powder can be "enhanced" with copper or other metal oxides as well as elements such as sodium, calcium, potassium etc.

The second heating process can be repeated three to four times, even applying different mixtures. When jewelry containing rubies is heated (for repairs) it should not be coated with boracic acid or any other substance, as this can etch the surface; it does not have to be "protected" like a diamond.

The treatment can identify by noting bubbles in cavities and fractures using a 10 x loupe.

Synthetic and Imitation Rubies

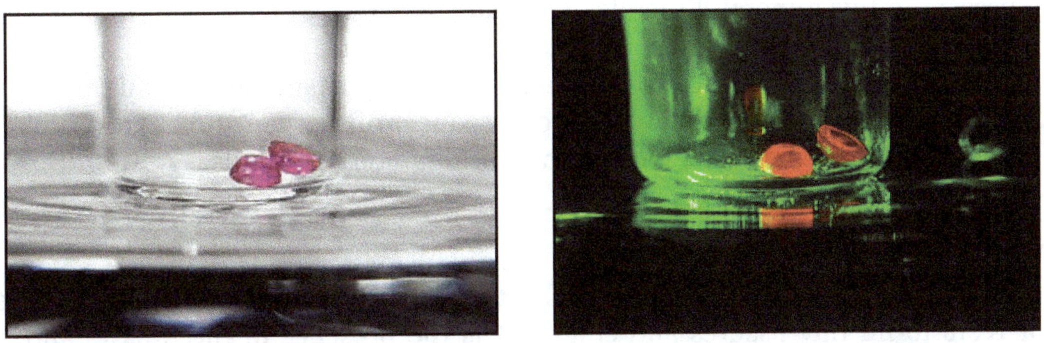

Artificial ruby under a normal light (top) and under a green laser light (bottom). Red light is emitted.

In 1837, Gaudin made the first synthetic rubies by fusing potash alum at a high temperature with a little chromium as a pigment. In 1847, Ebelmen made white sapphire by fusing alumina in boric acid. In 1877, Frenic and Freil made crystal corundum from which small stones could be cut. Frimy and Auguste Verneuil manufactured artificial ruby by fusing BaF_2 and Al_2O_3 with a little chromium at red heat. In 1903, Verneuil announced he could produce synthetic rubies on a commercial scale using this flame fusion process. By 1910, Verneuil's laboratory had expanded into a 30 furnace production facility, with annual gemstone production having reached 1,000 kilograms (2,000 lb) in 1907.

Other processes in which synthetic rubies can be produced are through Czochralski's pulling process, flux process, and the hydrothermal process. Most synthetic rubies originate from flame fusion, due to the low costs involved. Synthetic rubies may have no imperfections visible to the naked eye but magnification may reveal curves, striae and gas bubbles. The fewer the number and the less obvious the imperfections, the more valuable the ruby is; unless there are no imperfections (i.e., a perfect ruby), in which case it will be suspected of being artificial. Dopants are added to some manufactured rubies so they can be identified as synthetic, but most need gemological testing to determine their origin.

Synthetic rubies have technological uses as well as gemological ones. Rods of synthetic ruby are used to make ruby lasersand masers. The first working laser was made by Theodore H. Maiman in 1960. Maiman used a solid-state light-pumped synthetic ruby to produce red laser light at a wavelength of 694 nanometers (nm). Ruby lasers are still in use. Rubies are also used in applications where high hardness is required such as at wear exposed locations in modern mechanical clockworks, or as scanning probe tips in a coordinate measuring machine.

Imitation rubies are also marketed. Red spinels, red garnets, and colored glass have been falsely claimed to be rubies. Imitations go back to Roman times and already in the 17th century techniques were developed to color foil red—by burning scarlet wool in the bottom part of the furnace—which was then placed under the imitation stone. Trade terms such as balas ruby for red spinel and rubellite for red tourmaline can mislead unsuspecting buyers. Such terms are therefore discouraged from use by many gemological associations such as the Laboratory Manual Harmonisation Committee (LMHC).

Emerald

Another one of the five cardinal gemstones, the emerald, is a variety of the mineral beryl. The main component of the mineral is $Be_3Al_2(SiO_3)_6$. Trace amounts of chromium and occasionally, vanadium is also present in this mineral. Emeralds are characterized by their green color. To be labeled as a top gem, an emerald must have a pure verdant green hue and also exhibit a high level of transparency. In ancient times, Egypt, India, and Austria were important emerald mining sites. Today, however, Colombia is regarded as the world's top producer of this gemstone. The Chalk Emerald, the Mogul Mughal Emerald, etc., are some of the world famous emeralds.

Identifying Emerald

Almost all natural emeralds contain distinct characteristic inclusions and almost all are treated with oil or resin to fill tiny fissures and cracks. It is the presence of these flaws and oil which makes it quite easy to identify and distinguish natural emeralds amongst other similar gemstones.

Artificial light will expose and amplify inclusions and fractures that prove the stone to be a natural emerald. Emerald color is owed to trace amounts of chromium and vanadium, and color is best admired under natural daylight. One of the easiest methods to identify green emeralds is to test for specific gravity (density) and hardness. Like all forms of beryl, emerald is harder than apatite, quartz and feldspar, but is slightly softer than spinel, topaz and sapphire. However, emerald is generally more fragile than other beryl, owing to its naturally included state.

Emerald Origin and Gemstone Sources

Emerald mining is almost exclusively done from host rocks, where the emerald has grown into small veins or on walls of cavities.

Colombia is the international center of emerald mining. The Muzo Mine, northwest of Bogotá, produces fine-quality stones of a deep green color. The Chivor Mine, northeast of Bogotá, is another important deposit, and other locations promise additional successful mining results.

Brazil has various deposits in Bahia, Goias and Minas Gerais. The stones are lighter than the Colombian ones; mostly yellow-green and are often free of inclusions. Brazil also supplies rare emerald cat's eyes and extremely rare emeralds with a six-ray star.

South African deposits are concentrated in the Northern Transvaal. But only five percent of the stones found in the Cobra and Somerset Mines are of good quality. Most stones are light or muddy and are only suitable for cabochon cuts.

Zimbabwe has several deposits of emerald; the most important is the Sandawana Mine in the south. The crystals are small, but of very good quality.

Russia has deposits in the Ural, north of Sverdlovsk. Good quality emeralds are rare; most stones are light or muddy and only suitable for cabochons.

Further deposits are in Afghanistan, Australia (New South Wales and Western Australia), Ghana, India, Madagascar, Malawi, Mozambique, Namibia, Nigeria, Pakistan, Tanzania, Zambia and the United States (North Carolina).

Emerald Color

With emeralds, even more so than other colored gems, it is the color that is the chief determinant of value. By definition, emeralds are a medium to darker green to blue-green and sometimes slightly yellowish-green beryl. Emeralds owe their fine green color to traces of chromium and vanadium impurities. The most popular and valuable color is a slightly bluish green in a medium dark tone with strong to vivid saturation, however, too much blue can decrease the value.

The term "Colombian emeralds" is often used to describe vivid, slightly bluish-green stones of a medium to medium dark color, regardless of their geographic origin. Emeralds of a lighter color are sometimes called "Brazilian emeralds", even if they were mined in Africa.

Emerald Clarity and Luster

Emeralds have a vitreous (glass-like) luster when cut and polished. Although their clarity can occur from translucent to opaque, transparent specimens are most desirable and are much more valuable. Clarity is important, but inclusions are tolerated more in emeralds than virtually any other gem. Unlike other beryl gems, emeralds often contain inclusions and other flaws. These flaws are not looked upon as negative attributes for emeralds, as they would be for other gemstones. Indeed, these flaws are considered part of the character of the stone and are used to assure the purchaser of a natural stone.

Emerald clarity is graded by eye, unlike diamond, where 10x loupe magnification is used to grade clarity. If an emerald has no inclusions that are visible to the naked eye, it is considered flawless. "Eye-clean" stones will command the highest prices, especially those with ideal color grade.

Emerald Cut and Shape

Emeralds are most often cut in a special cut designed just for this gem; "the emerald-cut". The emerald cut is a step-cut or trap-cut featuring a rectangular or square shape with truncated corners. This cutting style maximizes the beauty and color of the stone, whilst protecting it from mechanical strain and internal stress. Emeralds are also cut into a variety of other traditional shapes such as pear, oval and round. Lower grade materials are often cut en cabochon or into beads. Highly transparent and clear materials are sometimes cut in brilliant-style.

Emerald Treatment

Oiling is a common emerald treatment and in most cases, it is done right at the mining location. The term ‹oiling' refers to the practice of immersing emeralds in a colorless oil or resin (most often cedar oil) in order to enhance color, clarity and stability. This is also often done using a vacuum chamber to assist penetration. Non-standard treatments go beyond this by using colored oils and epoxy-like resins.

These treatments dramatically improve the appearance of the gems, but necessitate special care in cleaning and setting. Steam cleaners, solvents and ultrasonic cleaners can remove the oils, making inclusions that were barely visible stand out in sharp relief. Since emeralds can be restored through re-oiling, the damage is considered to be only temporary.

Synthetic emeralds have been available since 1848. In the 1950s, synthesizing methods became commercialized and excellent quality synthetics have been made available since. The term 'Chatham emeralds' is often used for lab-grown emeralds. There are also numerous emerald doublets available. In many cases, a natural upper-half is cemented to a glass or synthetic pavilion using an emerald-green paste. With the visibility of emerald flaws from the crown down, once these are set into jewelry, they are often sold as natural emeralds.

Emerald Gemological Properties

Species:	Emerald - beryl
Chemical Formula:	$Al_2Be_3Si_6O_{18}$ - Aluminum beryllium silicate
Crystal Structure:	Hexagonal; hexagonal prisms
Color:	Emerald green to green to slightly bluish or yellowish-green
Hardness:	7.5 to 8 on the Mohs scale

Refractive Index:	1.565 to 1.602
Density:	2.67 to 2.78
Cleavage:	Indistinct
Transparency:	Transparent to opaque
Double Refraction or Birefringence:	-0.006
Luster:	Vitreous
Fluorescence:	Usually none

Synthetic Emerald

Emerald showing its hexagonal structure.

Both hydrothermal and *flux-growth* synthetics have been produced, and a method has been developed for producing an emerald overgrowth on colorless beryl. The first commercially successful emerald synthesis process was that of Carroll Chatham, likely involving a lithium vanadate flux process, as Chatham's emeralds do not have any water and contain traces of vanadate, molybdenum and vanadium. The other large producer of flux emeralds was Pierre Gilson Sr., whose products have been on the market since 1964. Gilson's emeralds are usually grown on natural colorless beryl seeds, which are coated on both sides. Growth occurs at the rate of 1 mm per month, a typical seven-month growth run produces emerald crystals 7 mm thick.

Hydrothermal synthetic emeralds have been attributed to IG Farben, Nacken, Tairus, and others, but the first satisfactory commercial product was that of Johann Lechleitner of Innsbruck, Austria, which appeared on the market in the 1960s. These stones were initially sold under the names "Emerita" and "Symeralds", and they were grown as a thin layer of emerald on top of natural colorless beryl stones. Later, from 1965 to 1970, the Linde Division of Union Carbide produced completely synthetic emeralds by hydrothermal synthesis. According to their patents (attributable to E.M. Flanigen), acidic conditions are essential to prevent the chromium (which is used as the colorant) from precipitating. Also, it is important that the silicon-containing nutrient be kept away from the other ingredients to prevent nucleation and confine growth to the seed crystals. Growth occurs by a diffusion-reaction process, assisted by convection. The largest producer of hydrothermal emeralds today is Tairus, which has succeeded in synthesizing emeralds with chemical composition similar to emeralds in alkaline deposits in Colombia, and whose products are thus known as "Colombian created emeralds" or "Tairus created emeralds". Luminescence in ultraviolet light is considered a supplementary test when making a natural versus synthetic determination, as many, but not all, natural emeralds are inert to ultraviolet light. Many synthetics are also UV inert.

Synthetic emeralds are often referred to as "created", as their chemical and gemological composition is the same as their natural counterparts. The U.S. Federal Trade Commission (FTC) has very strict regulations as to what can and what cannot be called a "synthetic" stone. The FTC says: "(c) It is unfair or deceptive to use the word "laboratory-grown," "laboratory-created," "manufacturer name-created," or "synthetic" with the name of any natural stone to describe any industry product unless such industry product has essentially the same optical, physical, and chemical properties as the stone named."

Amethyst

Its color is as unique as it is seductive, though in fact this gemstone of all gemstones is said to protect its wearer against seduction.

The amethyst is extravagance in violet. For many thousands of years, the most striking representative of the quartz family has been a jewel coveted by princes both ecclesiastical and secular. Moses described it as a symbol of the Spirit of God in the official robes of the High Priest of the Jews, and the Russian Empress Catherine the Great sent thousands of miners into the Urals to look for it. In popular belief, the amethyst offers protection against drunkenness - for the Greek words 'amethystos' mean 'not intoxicated' in translation. A more apt stone for the month of February, particularly if there is to be plenty going on in the way of carnival celebrations, could thus hardly be wished for.

A large number of further miraculous powers are attributed to the amethyst in all sorts of cultures. It was said to protect crops against tempests and locusts, bring good fortune in war and in the hunt, drive out evil spirits and inspire the intellect. A little study of the works of Pliny will reveal that this gemstone, if worn round the neck on a cord made from dog's hair, affords protection against snakebite. Later, Hieronymus even reported that eagles placed an amethyst in their nest in order to protect their young from the selfsame danger. Apart from these powers, gemstone therapists say that the amethyst has a sobering and cleansing effect.

Amethyst has also been said to quell excessive stomach acid and, served to combat insect bites and beautify the skin. But the amethyst not only had a firm niche in medicine; it was also esteemed as a stone of friendship. And since it was thought to put the wearer in a chaste frame of mind and symbolize trust and piety, the amethyst came to occupy a very prominent position in the ornaments of the Catholic clergy over the centuries. It was the stone of bishops and cardinals; we find it in prelates' crosses and in the so-called Papal Ring (Italian, 15th century) in the Jewelry Museum in Pforzheim.

However, the most beautiful of all crystal quartzes also posed one or two riddles for the scientists, and indeed they still haven't been completely solved to this day. The amethyst has its hardness, its moderate refraction and its weight in common with the other quartzes, but the crystal structure is different, and it is most unconventional. The construction is stratified, as a result of which areas and lamellae of varying color intensity often come about. This explains why there are relatively few large cut amethysts of an evenly distributed dark color, in spite of its having been found so abundantly in all parts of the world. It is only in the last few years that scientists have been fairly certain of having found the real cause of the color. It is now attributed to certain iron constituents in connection with natural radioactive radiation.

One thing that has been known for a long time, on the other hand, is the fact that the amethyst changes its color on being heated. Smoky stones are transformed at temperatures of as little as 250 degrees to a shining yellow to brownish-red, whilst clear ones, i.e. those with a high degree of transparency, become yellow or colorless at 400 degrees. Now and then Nature gives us a surprise by having created bicolored stones, like the ones recently found in Bolivia in the form of causticised crystal nuggets. This variety is known as ametrine, for in its formation certain energy states of iron introduce violet areas to the yellow citrine. At best, flat jewelry with a three-spoked star can be cut from it. However, the highlight for esoterics is that the energy fields can in fact be made visible in polarised light. The Henn Brothers of Idar-Oberstein even supply the photos to go with it.

Some amethysts pale almost to colourlessness in daylight. The reason for this has not yet been discovered, but it is possible to re-colour them by means of radium radiation. The fact that these stones can lose their colour makes it obvious that amethyst jewellery should not be worn while sunbathing, in a solarium or in a discotheque with black light. Sudden changes of temperature can also be harmful to the stone.

Asterism

Asterism" is a term, which is most commonly used in gemology. It refers a kind of special optic phenomenon that has been seen in several gems.

Special optic phenomena refer uncommon and excellent optic properties exhibit by some rare gemstone. The phenomena are particular property of transmitting or reflecting light. The structure or inclusions of a mineral in gems is the reason of this property. The beauty and value of gems depend on their optical properties.

"Asterism"-one of the important optical phenomena has been seen in sapphires, rubies, rose quartz and some other gems. The term "Asterism" came from the Greek word "aster", which means star. From the name we can guess "asterism" is star-shaped light effect across the surface of gems.

Special Optical Phenomenon

"Asterism" is an exclusive light effect (which we called special optical phenomenon) of some crystal. When light pass through these crystals, its exhibit a star-shaped figure, which called "asterism". This figure is produced by reflected or transmitted light.

Asterism in Gems

Gems, which show an illuminated star, just bellow the surface of crystal called star gems. The most popular star gems are star ruby and star sapphire. Other gems, which also possess the special optic phenomenon "asterism", are rose quartz, garnet, spinel, Chrysoberyl and Diopside. There are a few specimen of rock crystal, which has the phenomenon.

The star in the star gems may have four to twelve rays depending on the crystal system of the gems. But the six- rayed star effect is most common. Usually, six-rayed star show by the star rubies and star sapphires. Sometimes these star gems exhibit the twelve-rayed star effect. But the twelve-rayed star effect in corundum is occasional. Rose quartz exhibits six-rayed star. Some garnets and spinel commonly have the star effect with four rays, but these star gems can show six-rayed star effect if cut in a distinct way. Some crystals like Chrysoberyl and Diopside only exhibit four-rayed star effect.

"Asterism" is caused by the dense inclusions of tiny fibres of foreign minerals in gems. Generally, these inclusions are the reason of the reflection of light and form a wavy star like formation, which moves around when the mineral is rotated. All the inclusions in the gems are formed via exsolution.

The inclusions must be long, very thin needle like which called as rutile. "*Asterism*" is produced when needles intersect each other in multiple groups. Small needles of rutile (titanium oxide), sometimes called "silk" are oriented inside the minerals in the direction of the crystal growth system. There is a relationship between the gem's growth and the number of rays produced by the reflected light.

There must be three or more chatoyant bands inside the gems to produce "asterism". The arrangement of needles in a hexagonal pattern produced six-rayed "asterism".

Exsolved inclusions form a pattern in the host mineral. The pattern may vary from mineral to mineral, but in case of specific mineral the pattern remain same. In case of corundum, we can see rutile is exsolved in three directions crossing at 60 degree or 120 degree in the basal and produce six or twelve-rayed star. When the rutiles crossed at 90 degree, it produces a four-rayed star. Star gems are translucent to opaque.The rays of star should cross at the top of these gems and be distinct.

Star Sapphire

The sharpness and the size of "asterism" in gems, which identified as phenomenon, depend on the quality and quantity of rutile needles. Small to smaller size of rutile needles help "asterism" to be sharper. When the needles are visible with open eyes as inclusions, the gems turned into a low translucent stone and lack "asterism".

Large quantities of rutile needles in a crystal produce phenomenal "aster". If the quantities of rutiles are small, "Asterism" is visible in only a small area. These gems are not recognized as phenomenal gems.The clarity of these gems can be improve by dissolving some rutile needles with heat treatment. Heat treatment allows various elements to be incorporated into crystal. Then rutile needles form by very gradual cooling system and induce "asterism".

Inclusions are not present in all star gems. Some specimens leave holes or empty areas in the host. Reflection creates from fibres or fibrous cavity in the crystal.

Types of Asterism

There are two types of "asterism", like Epiasterism and Diasterism. Epiasterism is creates by the reflection of light on parallel-arranged inclusions inside the crystal. Diasterism is creates by the transmitting light through the stone, mostly seen in rose quartz. To see this effect a lighted background is essential.

To display the star effect in best way a gem is cut as cabochon with the girdle parallel to the needles. The domed stones are round or oval shaped.

The special optical phenomenon "asterism" is most visible in direct, single pointed light source, like an incandescent bulb, fibre optic light, and penlight or in sunlight.

The distinctiveness of the star and degree of strait ness of each ray help to judge the quality and value of star gems.

Chatoyance

Chatoyance is an optical phenomenon in which a band of reflected light, known as a "cat's-eye," moves just beneath the surface of a cabochon-cut gemstone. Chrysoberyl and tiger's-eye are two of the best-known gem materials that exhibit this phenomenon. Excellent specimens of chrysoberyl exhibit the finest chatoyance, and tiger's-eye is the chatoyant gem most widely used in jewelry.

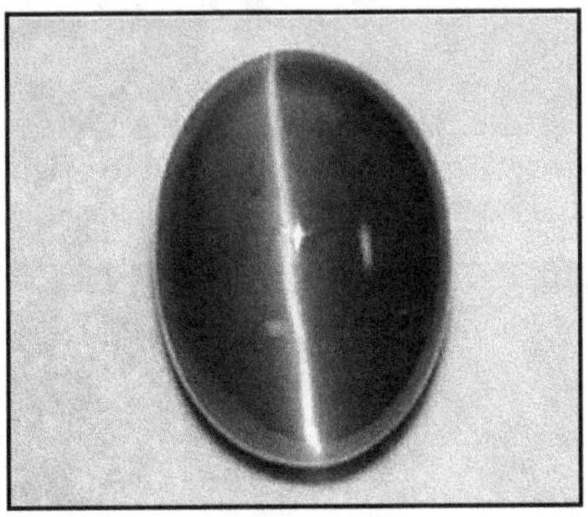

Cat's-eye chrysoberyl: Chrysoberyl is the gem that exhibits the finest "cat's-eye." When the name "cat's-eye" is used alone, it is understood that the speaker is referring to a cat's-eye chrysoberyl. When the speaker is referring to any other gem with a cat's-eye, the name of that gem is used after the words "cat's-eye." For example: "cat's-eye apatite." This green cat's-eye chrysoberyl was produced in Sri Lanka and is about 5.6 x 4 millimeters in size.

Chatoyance occurs in stones that contain a large number of very thin parallel inclusions within the stone, known as a "silk." The light reflects from these inclusions to form a thin band across the surface of the stone. The band of light always occurs at right angles to the length of the parallel inclusions. These inclusions can be crystals, hollow tubes, or other linear structures that are present throughout the stone and are usually aligned with a crystallographic axis. Needle-like crystals of rutile and hematite are well-known for producing a cat's-eye in many specimens.

Tiger's-eye cabochons: Two "honey" tiger's-eye cabochons with very nice chatoyance. These stones are 8 x 6 millimeter ovals and cost a few dollars each. They were cut from material mined in the Central African Republic. Tiger's-eye is the only chatoyant stone that is produced in large quantities in calibrated sizes. That makes it the only chatoyant stone commonly seen in mass-production jewelry stores. It is very popular in men's rings and cufflinks.

Chatoyance - Eye of the Cat

The name "chatoyance" originates from the French word "*chatoyer*," which means "to shine like a cat's eye." The analogy also matches the way a cat's pupils will narrow to a thin slit under bright light.

Cat's eye with a slit pupil: This cat's eyes have responded to bright light by closing their pupils to a narrow slit. This is where the "cat's-eye" gemstone name originates.

Cat's-eye actinolite: Two cat's-eye actinolite cabochons cut from material mined in Tanzania. Actinolite is an amphibole mineral that often has a fibrous internal texture that serves as a "silk" and produces very strong eyes. These stones are 7-millimeter rounds.

Understanding Chatoyance

A good analogy for the chatoyance effect is how light reflects to form a line across the surface of a spool of silk thread. Viewing and moving a spool of silk thread under a beam of incident light can be a useful way to develop an understanding of how parallel inclusions produce an "eye" within a gemstone.

Cat's-eye on silk thread: These spools of silk thread exhibit chatoyance. Light shining on the spools reflects from the lustrous silk threads. As a result, a line of light similar to the "eye" of a gemstone forms across the portion of the spool where the tangent of the spool's surface is at a right angle to the observer's line of sight.

In a chatoyant gemstone, the band of light will move back and forth beneath the surface of the gem as it is turned under a beam of incident light. The band will also move if the position of the light is moved, or the observer moves his head to view the stone from a different angle. The motion of the cat's-eye across the top of the gem is one of the things that makes these stones so interesting, beautiful, and desired by many people.

Cat's-eye apatite: Apatite is a mineral that often produces a very nice cat's-eye on a gemmy base color. The golden-green cabochon on the left was cut from material mined in Kenya. It is a 7 x 6 millimeter semitransparent oval with a nice eye and a great base color. The golden-brown stone on the right was also cut from material mined in Kenya. It is a 9 x 7 millimeter translucent oval.

Many Types of Gems Exhibit Chatoyance

Many types of gems exhibit chatoyance; however, it is not present in every specimen of those gems and is usually only seen in a minority of specimens. Gems with well-known chatoyance include the following:chrysoberyl, tiger's-eye, actinolite, apatite, beryl (aquamarine, heliodor, emerald), beryllonite, cerussite, danburite, diaspore, diopside, enstatite, garnet, iolite, kyanite, moonstone, peridot, pezzottaite, prehnite, quartz, rutile, sillimanite, scapolite, spinel, topaz, tourmaline, and zircon.

Cat's-eye sillimanite: This cabochon is an 11 x 8 millimeter oval cut from a purplish brown sillimanite mined in Madagascar. It has a distinct pink eye on a background that is nearly opaque.

Cat's-eye aquamarine: Rare translucent aquamarines are known to produce cat's-eyes, although the eyes are generally not very strong. This greenish blue stone is about 11 x 8 millimeters in size and was cut from material mined in Madagascar.

Milk and Honey

Some gems with a highly developed chatoyance can appear to be made of two different materials when illuminated from the proper direction with respect to the observer's eye. In these stones, the cat's-eye will appear to divide the stone into a zone of light-colored material on one side of the eye and dark-colored material on the other. This phenomenon is known as the "milk-and-honey" effect. A photo of cat's-eye chrysoberyl showing the milk-and-honey effect is included below.

Milk-and-honey chrysoberyl: Chrysoberyl is the gem that exhibits the finest cat's-eye. When the term "cat's-eye" is used without a mineral name as a modifier, the speaker is most likely referring to chrysoberyl. This specimen exhibits the "milk-and-honey" effect - when properly oriented, the stone has two apparently different colors on each side of the cat's-eye line. This green cat's-eye chrysoberyl was produced in Sri Lanka and is about 5.6 x 4 millimeters in size. The stone in this photo is the same stone shown in the photo at the top of the page.

Cat's-eye heliodor: This golden-yellow heliodor beryl was mined in Madagascar and cut into a 10 x 8 millimeter oval. It has a beautiful translucent color and a faint eye.

Cutting the Cat's-Eye

Anyone who cuts cabochons with the intent of producing a cat's-eye stone must first find a material with a silk of inclusions that is capable of producing an eye. Then the cutter must examine the rough and orient the stone so that the silk will be parallel to the bottom of the stone within the finished cabochon, and the parallel inclusions will be parallel to the long axis of the cabochon. If this proper orientation is not achieved, the stone will have an off-center eye or no eye at all.

Cat's-eye tourmaline: Tourmaline often contains a silk of coarse tubes that can produce a very strong cat's-eye. In these stones you can see how the silk crosses the stone at right angles to the cat's-eye. A skilled cutter must assess the stone before cutting so that this orientation can be achieved. These colorful cabochons were cut from material mined in Brazil. The pink stone is an 8.5 millimeter round, and the green stone is a 10 x 7 millimeter oval.

Assessing the Quality of a Cat's-Eye Gem

The best cat's-eye gems have an eye that meets the following criteria:

- It is clearly visible.
- It symmetrically bisects the cabochon.

- It contrasts sharply with the stone's body color.
- It moves smoothly as the stone is turned under light.

If a stone features all of the above criteria and has an exceptional body color, then you have a fantastic cat's-eye gem. The importance of a desirable body color cannot be overstated.

Cat's-eye scapolite: This scapolite cabochon, cut from material mined in India, is especially interesting. It contains a very coarse silk that not only produces a cat's-eye, but also serves as a diffraction grating that produces a strong display of spectral colors. The stone is a 12 x 9 millimeter oval.

Diffraction in Cat's-Eye

Rare chatoyant specimens will have a coarse silk with just the right spacing to serve as a diffraction grating. These specimens will produce not only a cat's-eye, but also a display of spectral colors caused by light passing through the coarse silk and being diffracted into the colors of the spectrum. More familiar examples of diffraction are rainbows produced by sunlight diffracted by raindrops, or opal's play-of-color caused by light diffracted by tiny spheres of opal within the gem. The specimen of cat's-eye scapolite in the photo on this page is an example. (Another stone that exhibits diffraction is iris agate).

References

- What-are-gemstones: eternaltools.com, Retrieved 17 May 2020
- Walker, J. "Optical absorption and luminescence in diamond". Rep. Prog. Phys. 42(10): 1605.doi:10.1088/0034-4885/42/10/001
- Gemology: investopedia com Retrieved 20 March 2020
- Martin Meredith (2007). Diamonds, Gold, and War: The British, the Boers, and the Making of South Africa. New York: Public affairs. p. 16. ISBN 1-58648-473-7
- The-five-cardinal-gemstones: worldatlas.com, Retrieved 08 April 2020
- Harman, Alang Kasim; Ninomiya, Susumu; Adachi, Sadao (1994). "Optical constants of sapphire (alpha-Al2O3) single crystals". Journal of Applied Physics. 76 (12): 8032–8036. Bibcode:1994JAP....76.8032H. doi:10.1063/1.357922

- Asterism-in-gems: jewelinfo4u.com, Retrieved 16, July 2020
- "A Contribution to the Understanding of Blue Fluorescence on the Appearance of Diamonds" (PDF). GIA. GIA. Winter. Retrieved 20, January 2020
- Diamond-gemstone: britannica.com, Retrieved 17, August 2020

Chapter 6

Crystallography and Minerals

Crystallography is the study of distinguishing the arrangement and bonding of atoms in crystalline solids and with the geometric structure of crystal lattices. Most minerals occur naturally as crystals and thus are studied under crystallography. Crystal symmetry, crystal systems, crystal classes, crystal lattices, etc. are some of its aspects. This chapter has been carefully written to provide an easy understanding of the subject matter and its related points.

Mineral Crystal

Crystals are usually associated with having naturally developed, flat and smooth external faces. It has long been recognised that this evidence of external regularity is related to the regularity of internal structure. Diffraction techniques are now available which give much more information about the internal structure of crystals, and it is recognized that internal order can exist with no external evidence for it.

2-D Crystallography

Repeating patterns can be described in terms of their symmetry. Similarly, the regularity of the internal structure of crystals can be described formally in terms of symmetry elements the symmetry and the categorization of patterns and crystals in just two dimensions to simplify the ideas presented. Two-dimensional crystals can be thought of as the projection of 3-D crystals onto a plane. Reflection symmetry is demonstrated and then rotation symmetry is explained. Exercise pages allow the user to identify reflection and rotation symmetry in 2-D shapes. The combination of reflection and rotation symmetry to give the ten plane point symmetry groups follow.

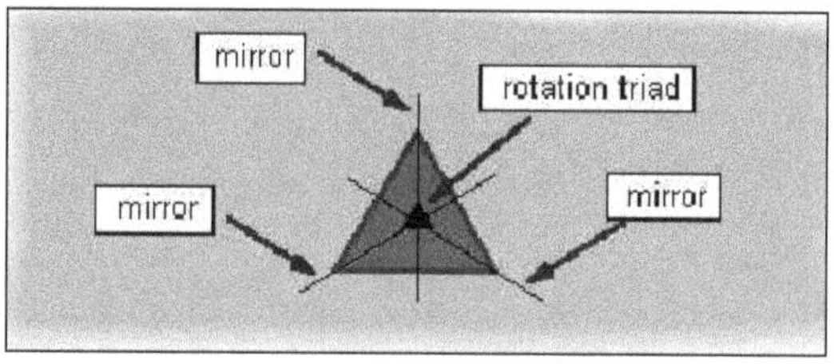

Two dimensional patterns and crystals can be built up from a single unit of pattern called a basis, by repeated translations, due to two vector operators, known as unit vectors. Unit vectors outline a repeated area of a pattern called a unit cell. Repeated translations of unit vectors mark out a grid of identical points in space, called a lattice. The definition of a lattice is very important and is given as: a repeating pattern of points, each point having the same surroundings. The concept that patterns and crystals can be built by associating an identical basis in the same way with each point of a lattice is demonstrated. Unit cells can be divided into primitive and non-primitive types. All 2-D lattices can be categorized into one of five classes of plane lattices. The unit cells for each of these five lattices are shown. By clicking on these unit cells, the user is shown the symmetry elements that are present in these five lattice types.

Finally, deals with translational symmetry elements, in this case the glide line. When all the symmetry types are combined with the five plane lattice type, then the 17 plane groups are created. All 2-D patterns and crystals can be assigned to one of these plane groups. The 17 groups are given and by clicking on the group name an example of a pattern is given. A flow chart follows which gives the user the opportunity to identify the plane group of patterns given in the additional question.

3-D Crystallography

All crystals can be assigned to be in one of seven crystal systems. Depending on the shape of their unit cells. Seven distinct unit cells exist; the shape of these unit cells is determined by the symmetry of the crystal system. Within each of the crystal systems, different lattices are possible. There are 14different lattices, known as the Brava is lattices.

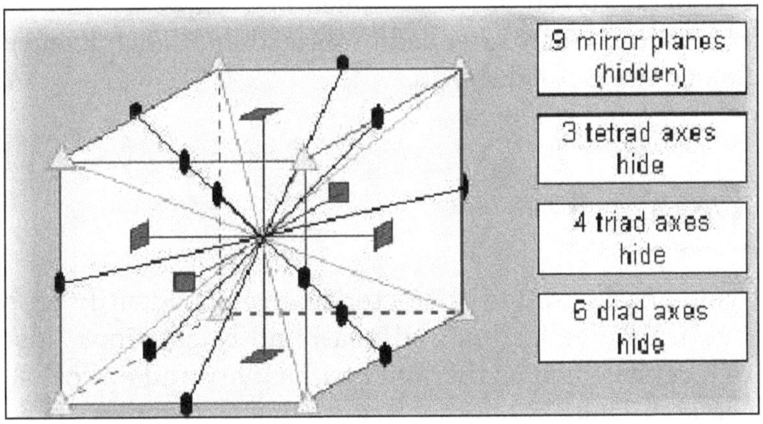

Mirrors lines become mirror planes, and objects can have more than one rotation axis in different directions. The se symmetry elements all pass through the center of the object, through a single point. This combination of symmetry elements is called the point symmetry group for an object.

Additional kinds of symmetry arise in three dimensions, which do not occur in two dimensions. These are centers of symmetry and inversion axes. These are demonstrated graphically. In total there are 32 different combinations of symmetry elements giving the 32 point symmetry groups. These point groups are distributed amongst the 7 crystal systems and are also known as the 32 crystal classes. If the 14 Brava is lattices and 32 point groups are combined and we consider the translational symmetry elements of glide planes and screw axes, then there are 230 different possible combinations. These 230 space groups describe all the possible different spatial arrangements of symmetry in crystals.

Crystal Structure

There are three common metallic crystal structures: face-centred cubic (fcc), hexagonal close-packed (hcp) and body-centred cubic (bcc). How these structures can be built by packing together atoms of the same size. Unit cells are presented for these three structures which can be rotated by the user. The user is instructed to try to find the rotation axes of these structures.

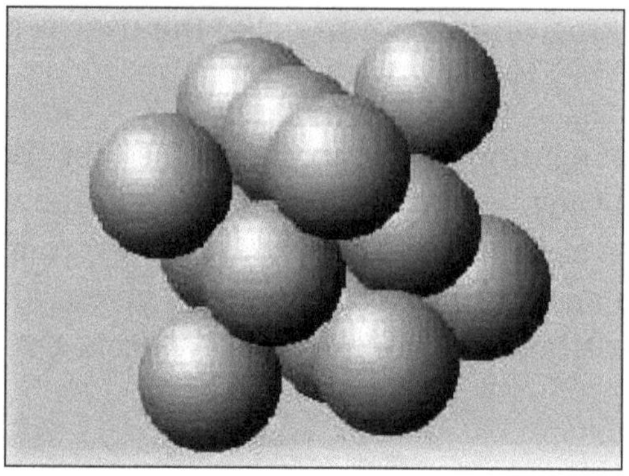

Indexing Directions and Planes

It is often necessary to state the direction in a crystal. A method is described which show show to calculate the Miller index for a direction in a 2-D lattice. It is important to be able to identify the unit vectors in the lattice of the crystal, since the Miller index expresses direction in terms of ratios of these vectors. This method is easily extended to three dimensions. Exercises allow the user to see if they have understood the method shown.

Crystal Symmetry

In crystallography, symmetry is used to characterize crystals, identify repeating parts of molecules, and simplify both data collection and nearly all calculations. Also, the symmetry of physical properties of a crystal such as thermal conductivity and optical activity must include the symmetry of the crystal. Thus, a thorough knowledge of symmetry is essential to a crystallographer.

An object is described as symmetric with respect to a transformation if the object appears to be in a state that is identical to its initial state, after the transformation. In crystallography, most types of symmetry can be described in terms of an apparent movement of the object such as some type of rotation or translation. The apparent movement is called the symmetry operation. The locations where the symmetry operations occur such as a rotation axis, a mirror plane, an inversion center, or a translation vector are described as symmetry elements.

Two distinct methods of describing rotational symmetry operations exist. These two sets of descriptors are the Hermann-Mauguin nomenclature and the Schönflies nomenclature. The Carl Hermann-Charles Mauguin system is typically used to describe crystals and crystallographic symmetry. The Arthur Schönflies convention is primarily used to describe symmetry in discrete molecules, in optical spectroscopy, and in quantum mechanics.

Crystals are defined as solids that have an atomic structure with long-range, 3-dimensional order. Unfortunately, this long-range order cannot be absolutely confirmed by any other method than some diffraction technique. However, there are several observations that can be made that will strongly suggest that a sample is crystalline before a diffraction experiment is undertaken.

Typically, crystals have flat faces and sharp edges. Also, many crystals will have one or more directions that can be cleaved cleanly. Samples with a naturally round shape or samples that have a concoidal fracture pattern are nearly always described as a glass having no significant, long-range, 3-D order. Similarly, materials that can be gently poked with a probe, and retain the deformed shape are gels or plastic materials and hence have not long-range, 3-D order.

When you look at several crystals from one material, you will soon notice that, although the crystals may have different sizes, all crystals have the same shape or habit. In particular, the angles between certain pairs of faces of the different crystals will be the same. This observation was first made by Nicholas Steno in 1669. This observation became known as the law of constancy of interfacial angles.

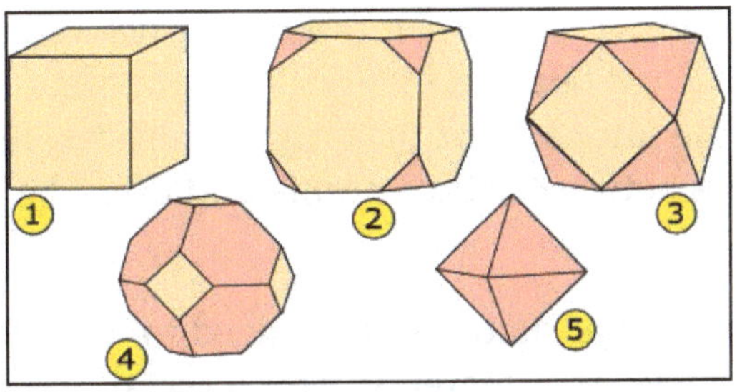

Figure: Models of crystals showing constant angles.

Steno and others in the 17th century were interested in the specific make up of crystals that would allow them to maintain the same angles between pairs of faces. These scientists believed that crystals must be made of some regularly-repeating components. Through these early studies, René-Just Haüy was able to postulate that if crystals of calcite and cubic garnets were built from many small blocks, then these blocks could easily be used to describe the faces of these crystals in terms of rational indices. This law of rational indices forms the basis of optical crystallography.

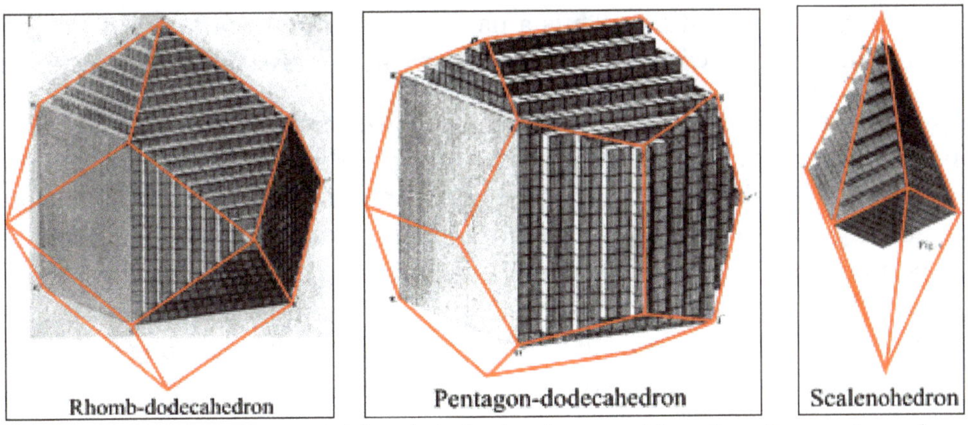

Figure: Models from Haüy's Traité de minéralogie - the crystal form have been redrawn in red.

Unit Cells

These regularly-repeating blocks are now known as unit cells. The dimensions of a unit cell are described by the lengths of the three axes, a, b, and c, and the three interaxial angles, α, β, and γ. In most published papers the axial lengths are expressed in terms of Å (Ångströms), and the interaxial angles are expressed in terms of ° (degrees).

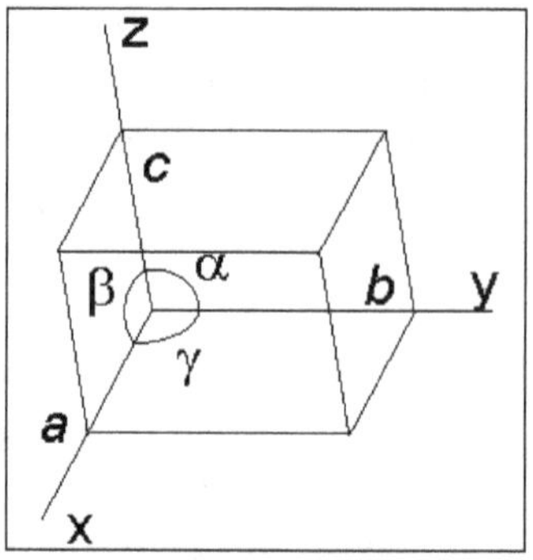

Figure: Unit cell showing cell parameters.

There are many choices of repeating blocks in any given lattice. The main principles defining the lattice is that each lattice point must be in an identical environment as any other lattice point, and that the individual blocks in the lattice must have the smallest volume possible. Often there are many ways to select the vectors between lattice points and even the locations of the lattice points themselves.

When researchers discuss a particular material, they need to work from one standard or conventional description of the unit cell for that material. Thus crystallographers have chosen the following criteria for selecting unit cells. By convention the unit cell edges are chosen to be *right-handed* ($a \times b$ is the direction of c), to have the *highest symmetry*, and to have the *smallest cell volume*. If other symmetry considerations do not override, then the cell is chosen so that $a \le b \le c$, and α, β, and γ all < 90 ° or all ≥ 90 °. This type of cell is called the *reduced* cell. There are several other rules for obtaining the conventional *reduced cell* for a given material.

Crystalline materials are separated into 7 crystal different systems. These crystal systems are most easily identified by the constraints on the cell parameters. However, that the cell parameter constraints are only necessary conditions. Thus a particular sample could have cell parameters that appear to fall into one category within experimental error, but are actually of lower symmetry.

The 7 crystal systems are listed in Table below. In the lowest symmetry system, triclinic, there are no restrictions on the values of the cell parameters. In the other crystal systems, symmetry reduces the number of unique lattice parameters as shown in the Table. Certain conventions have been followed in tabulating the parameters. In the monoclinic system, one of the axes is unique in the sense that it is perpendicular to the other two axes. This axis is selected by convention as either the *b* or *c* axis so that either β or γ are ≥ 90°, respectively. That *c*-unique monoclinic cells are common in French literature and *b*-unique cells are common in most other languages. In the tetragonal, trigonal, and hexagonal systems, one axis contains higher symmetry. By convention this axis is selected as the *c* axis.

Table: Crystal Systems.

Crystal System	#	Cell Parameters	Symmetry
Triclinic	6	$a \ne b \ne c; \alpha \ne \beta \ne \gamma$	$\bar{1}$
Monoclinic	4	$a \ne b \ne c; \alpha = \gamma = 90°, \beta \ge 90°$	2/m
Orthorhombic	3	$a \ne b \ne c; \alpha = \beta = \gamma = 90°$	mmm
Tetragonal	2	$a = b \ne c; \alpha = \beta = \gamma = 90°$	4/m, 4/mmm

Trigonal			
Hexagonal	2	$a = b \neq c; \alpha = \beta = 90°, \gamma = 120°$	$\bar{3}$, $\bar{3}$m
Rhombohedral	2	$a = b = c; \alpha = \beta = \gamma \neq 90°$	$\bar{3}$, $\bar{3}$m
Hexagonal	2	$a = b \neq c; \alpha = \beta = 90°, \gamma = 120°$	6/m, 6/mmm
Cubic	1	$a = b = c; \alpha = \beta = \gamma = 90°$	2/m$\bar{3}$, m$\bar{3}$m

The seven crystal systems each describe separate ways that simple 3-dimensional lattices may be constructed. As with all lattice systems, crystalline lattices are considered to have lattice points on the corners of the unit cell. Lattice points are selected so that the local environment around any particular lattice point is identical to the environment around any other lattice point.

Some trigonal lattices can be expressed on the basis of either a hexagonal or rhombohedral lattice. These lattices are shown in the drawings below. That rhombohedral lattice vectors can be expressed in an obverse orientation a), or in a reverse orientation b) below.

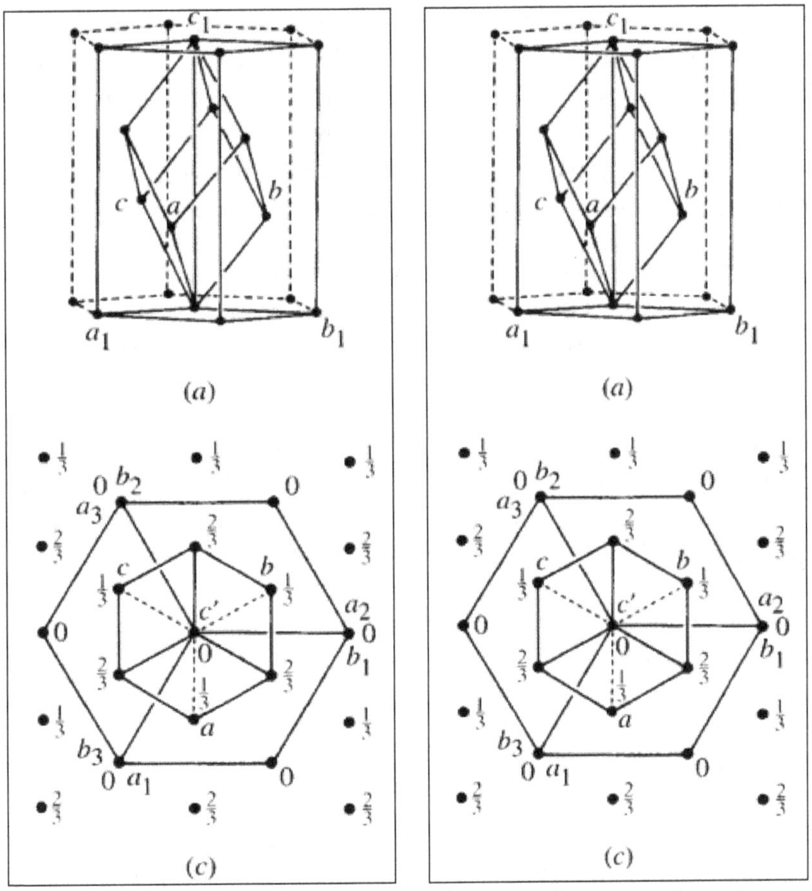

Figure: Rhombohedral lattices in the obverse and reverse orientations.

Bravais Lattices

It is sometimes possible to generate a lattice with higher symmetry if the lattice vectors are chosen so that one or more lattice points are also on the center of a face of the lattice or inside of the unit cell. These lattices with additional lattice points are described as centered lattices. Lattices with lattice points only on the corners are called primitive and are designated with the symbol P. That reduced cells described above are always primitive. In 1849, Auguste Bravais found that all regular crystals could be described in terms of only 14 lattice types for the 7 crystal systems. Remember that the environment around any one lattice point is exactly the same as the environment around any other lattice point.

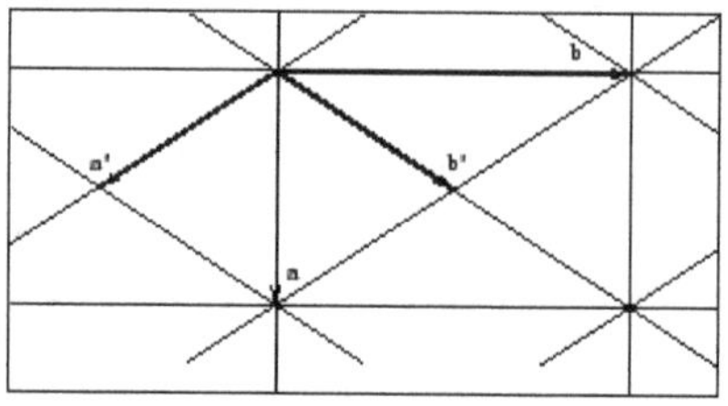

Figure: A c-centered lattice showing the a and b lattice vectors and the a' and b' vectors of the primitive cell.

Non-primitive lattices can have one, two, or three additional lattice points per unit cell. Lattices with one additional lattice point such as an A-, B-, or C-centered cells have an additional lattice point located on the center of some face of the crystal. The additional lattice points in A-centered cells appear on the b-c faces. Similarly, additional lattice points in B-centered cells appear on the a-c faces, and lattice points will appear on the a-b faces of C-centered cells. Lattices with body-centered, I-centered, cells have an additional lattice point in the center of the cell. Rhombohedral cells that are based on a hexagonal lattice conventionally have lattice points at (2/3, 1/3, 1/3) and (1/3, 2/3, 2/3). Face-centered, F-centered cells have lattice points on all faces.

Generally higher metric symmetry is identified by computer programs. However, these centered lattices can sometimes be identified by a simple examination of the cell parameters. When two cell lengths of a reduced lattice are approximately equal and the two corresponging cell angles are also approximately equal, then the cell is probably centered.

The following table lists the 14 Bravais lattice types. The Bravais symbols are a combination of the crystal system and the lattice designation. Triclinic types begin with the letter a that stands for anorthic from the mineral anorthite a mineral found to have triclinic symmetry. The other lattice types generally begin with the first letter of the crystal system.

Table: Bravais Lattice Symmetry.

Crystal System	Bravais Lattice	Laue Symmetry
Triclinic	aP	$\bar{1}(C_i)$
Monoclinic	mP, mS	$2/m(C_{2h})$
Orthorhombic	oP, oS, oF, oI	$mmm(D_{2h})$
Tetragonal	tP, tI	$4/m(C_{4h})$, $4/mmm(D_{4h})$
Trigonal		
Hexagonal	hP	$\bar{3}(C_{3i})$, $\bar{3}m(D_{3d})$
Rhombohedral	rR	$\bar{3}(C_{3i})$, $\bar{3}m(D_{3d})$
Hexagonal	hP	$6/m(C_{6h})$, $6/mmm(D_{6h})$
Cubic	cP, cF, cI	$m\bar{3}(T_h)$, $m\bar{3}m(O_h)$

The *S* symbol for monoclinic lattices represents a lattice with *A, C,* or *I* centering (*b*-unique) or *A, B,* or *I* centering (*c*-unique). The *S* symbol for orthorhombic lattices stands for any of the three side-centered lattice types, *A, B,* or *C.* Since *P* trigonal lattice and a *P* hexagonal lattice are identical in appearance, these two systems are considered to make up only one Bravais lattice type.

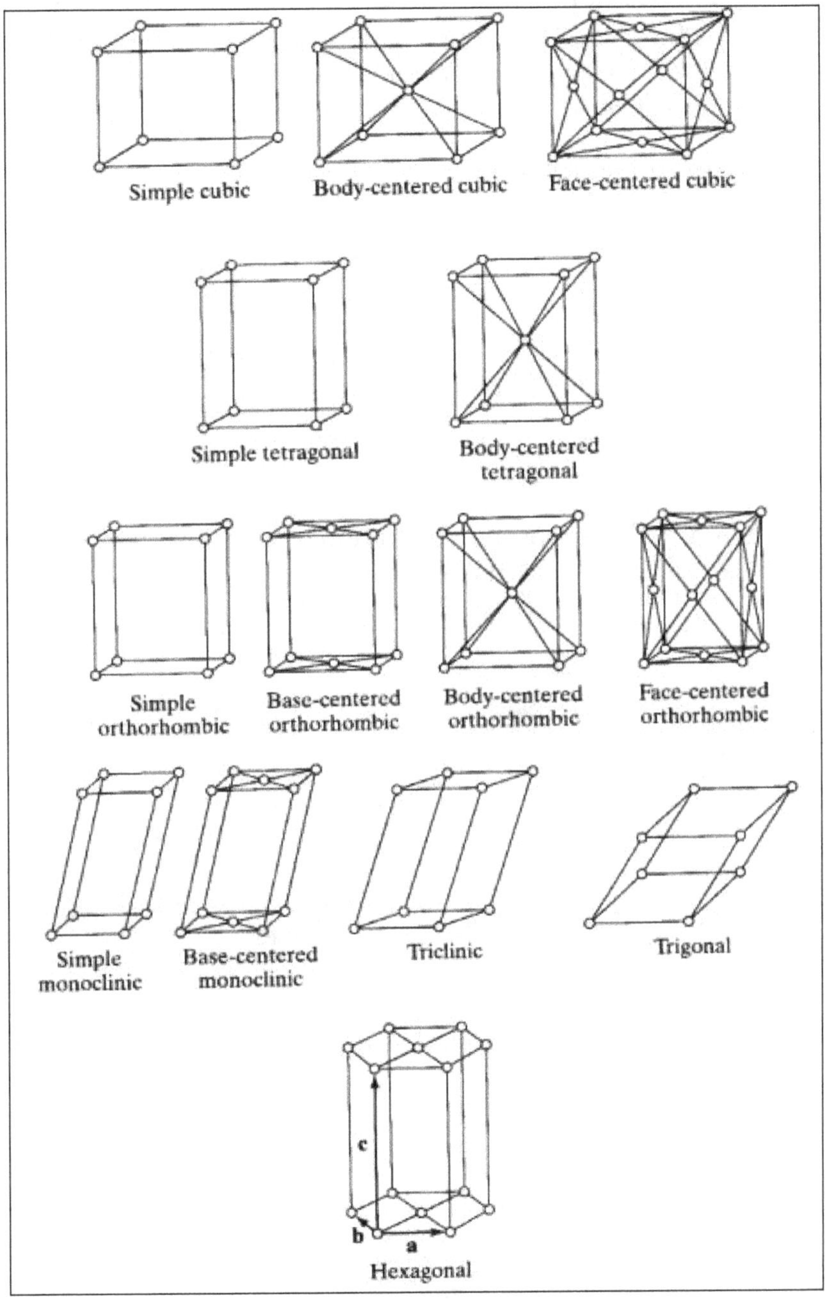

Figure: Bravais Lattices.

Rotational Symmetry

There are two basic types of rotational symmetry operations. *Proper* rotations move an object, but do not change the handedness of the object. *Improper* rotations include a proper rotation as well as a component that inverts the handedness of the object.

Proper Rotation

An n-fold (C_n) proper rotation operation represents a counter-clockwise movement of (360/n)° around an axis through the object. If an n-fold rotation operation is repeated n times, then the object returns to its original position. Crystals with a periodic lattice can only have axes with 1-, 2-, 3-, 4-, and 6-fold symmetry axes. The following description of rotational symmetry operations is similar to that given by Prof. Stephen Nelson. In the following drawings; the symmetry axis extends perpendicular from the page:

- 1-Fold Rotation: A 1(E)-fold rotation operation implies either a 0° rotation or a 360° rotation, and is referred to as the *identity* operation.

- 2-Fold Rotation: A 2-fold ($C2$) rotation operation moves the object by (360/2)° = 180°. The symbol used to designate a 2-fold axis is a solid oval.

- 3-Fold Rotation: A 3-fold ($C3$) rotation operation moves the object by (360/3)° = 120°. The symbol used to designate a 3-fold axis is a solid equilateral triangle.

- 4-Fold Rotation: A 4-fold ($C4$) rotation operation moves the object by (360/4)° = 90°. The symbol used to designate a 4-fold axis is a solid square.

- 6-Fold Rotation: A 6-fold (*C6*) rotation operation moves the object by (360/6)° = 60°. The symbol used to designate a 6-fold axis is a solid hexagon.

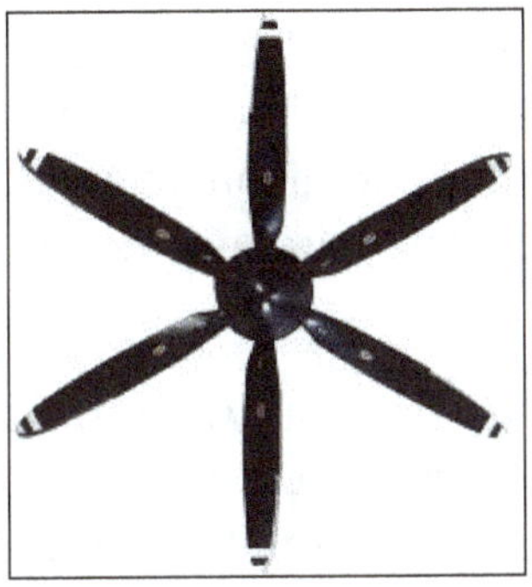

Improper Rotation

An improper rotation may be thought of as occurring in two parts, first a proper rotation is performed, followed by an inversion through a particular point on the rotation axis. In the H-M nomenclature, improper rotations are sometimes called roto-inversions. In the Schönflies scheme, improper rotations are roto-reflection axes because they are a rotation followed by a reflection perpendicular to the rotation axis. Improper rotations are designated by the symbol $\bar{n}$, where n represents the type of proper rotation component of the operation. As in the proper rotation operations, only $\bar{1}$ ($i = S_2$), $m = \bar{2}$ ($\sigma = S_1$), $\bar{3}$ (S_6), $\bar{4}$ (S_4), and $\bar{6}$ (S_3) improper rotations are commonly observed in crystals. These axes are pronounced as 3 bars in the United States and bar 3 in many European countries. Thus $\bar{3}$ in H-M is equivalent to S_6 in Schönflies.

That it is not necessary for either the rotation operation or the inversion center to exist as an operation of the group for the improper rotation axis to exist, e.g. the $\bar{4}$ (S_4) operations contains neither a 4-fold rotation axis (C_4) nor an inversion center:

- $\bar{3}$ Roto-inversion: This operation involves a rotation by (360/3) ° followed by an inversion through the center of the object. The symbol is a filled triangle with an open circle in the middle. This is the only improper rotation that also includes the proper rotation axis and an inversion center.
- $\bar{4}$ Roto-inversion: This operation involves a rotation by (360/4) ° followed by an inversion through the center of the object. The symbol is an open 4-sided diamond with an filled oval in the middle.
- $\bar{6}$ Roto-inversion: This operation involves a rotation by (360/6) ° followed by an inversion through the center of the object. The symbol is an open hexagon with a filled triangle inside.

Stereographic Projection

Drawing the three-dimensional symmetry operations on a two-dimensional surface such as this page has been a difficult problem. One way to overcome this problem is through the use of a stereographic projection. Such figures are also effective in describing the angular relations among the faces of a crystal.

To construct a stereographic projection, imagine that the object with a given symmetry or surface is at the center of a sphere. Consider the sphere to have a polar axis that is bisected by an equatorial plane. Project features of interest on the object from the center, out to the surface of the sphere. Then project the points on the surface of the sphere through the equatorial plane to the point where the polar axis intersects the sphere in the opposite hemisphere. The stereographic projection is then given by the equatorial plane and all intersections of the plane by the projected points. If the projection point started in the northern hemisphere then its projection onto the equatorial plane is represented as a plus. Points originating in the southern hemisphere are denoted with a circle. Sometimes points generated by improper symmetry operations are also denoted with a comma to indicate opposite handedness.

The unit cell axis with highest symmetry is usually selected as the polar axis. Rotation axes not in the equatorial plane are drawn with the symbol representing the type of axis at the projection point on the equatorial plane. Rotation axes in the equatorial plane are drawn outside of the projection terminating in arrows. Mirror planes are drawn as thickened lines. Inversion centers are drawn as open circles in the center of the polar axis.

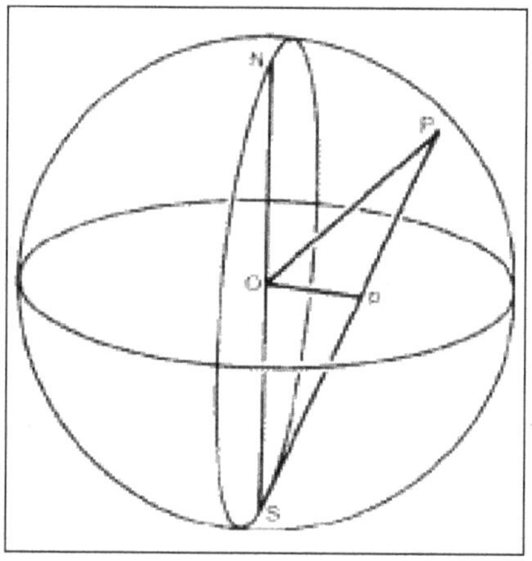

Crystallographic Point Groups

Symmetry operations can be combined to generate other symmetry operations. When these operations are written mathematically, the operations are applied in a right-to-left order. Thus in the expression below, the object x is operated on first by C and then by B. The operation symbol * only indicates that some transformation is being applied to the object.

$$A * x = B * C * x$$

These operations can be thus combined to form a *group* of symmetry operations. These groups of operations are called *point groups* because the symmetry elements of these operations all pass through a single point of the object. In the study of groups, mathematicians have found that such groups always have the following properties:

- *Identity:* One operation of the group must exist that when operated either before or after any other group operation produces the same transformation as that of the other group operation. Sometimes the identity operation is called the «do nothing" operation. (A * E = E * A = A, E is the identity operation).
- *Inverse:* For each operation of the group there must exist a second operation that when combined with the first operation produces the identity operation. The inverse operation cancels the effect of any operation. The identity operation is also its own inverse. (A * B = B * A = E, E is the identity, B is the inverse of A).
- *Associativity:* The order of combining the operations does not matter. In addition to these properties, all crystallographic symmetry groups possess the next property.

  [A * (B * C) = (A * B) * C].
- *Closure:* When any two operations of the group are combined, then the resultant operation must be a member of the group. (A = B * C, A, B, C are operations of the group).

Finally, some groups also exhibit the property of *commutativity*. The order of operations in commutative groups does not matter. (A * B = B * A), When the proper and improper rotation operations are combined following the rules of groups, they yield a total of 32 unique crystallographic point groups. These groups are listed in the following table. The centrosymmetric point groups are shown in bold.

Table: Crystallographic Point Groups.

System	Essential Symmetry	Point Groups
Triclinic	none	1, $\mathbf{\bar{1}}$
Monoclinic	2 or *m*	1, *m*, **2/*m***
Orthorhombic	222 or *mm*2*	222, *mm*2, ***mmm***
Tetragonal	4 or $\bar{4}$	4, 422, $\bar{4}$, **4/*m***, 4*mm*, $\bar{4}$2*m*, **4/*mmm***
Trigonal	3 or $\bar{3}$	3, $\mathbf{\bar{3}}$, 32, 3*m*, $\mathbf{\bar{3}}$***m***
Hexagonal	6 or $\bar{6}$	6, 622, $\bar{6}$, **6/*m***, 6*mm*, $\bar{6}$2*m*, **6/*mmm***
Cubic	23	23, **2/*m*-3**, 432, $\bar{4}$ 3 m, **4/*m*$\bar{3}$2/*m* = *m*$\bar{3}$*m***

By convention the following rules have been adopted to describe point groups. When a rotation axis is followed by a slash and an m, then this mirror is perpendicular to the rotation axis. For orthorhombic systems the three characters describe the symmetry along the three axes, a, b, and c, respectively. For tetragonal, trigonal, and hexagonal type cells, the c axis is unique, and the first symbol in the point group shows the symmetry along the unique axis. In tetragonal systems, the second symbol shows the symmetry along the [100] and [010] directions and the third symbol shows the symmetry along the [110] and $[1\bar{1}0]$ directions. In trigonal and hexagonal cells, the second symbol shows the symmetry along [100], [010] and $[\bar{1}\bar{1}0]$, and the third symbol shows symmetry along [210], [120], and $[\bar{1}\bar{2}0]$. In rhombohedral systems on rhombohedral axes, the first symbol shows symmetry along [111], and the second symbol shows symmetry along $[1\bar{1}0]$, $[01\bar{1}]$, and $[\bar{1}01]$. Cubic symbols show [100], [010], [001] in the first symbol, [111], $[1\bar{1}\bar{1}]$, $[\bar{1}1\bar{1}]$, $[\bar{1}\bar{1}1]$ in the second symbol and $[1\bar{1}0]$, [110], $[01\bar{1}]$, [011], $[\bar{1}01]$, and [101] in the third symbol.

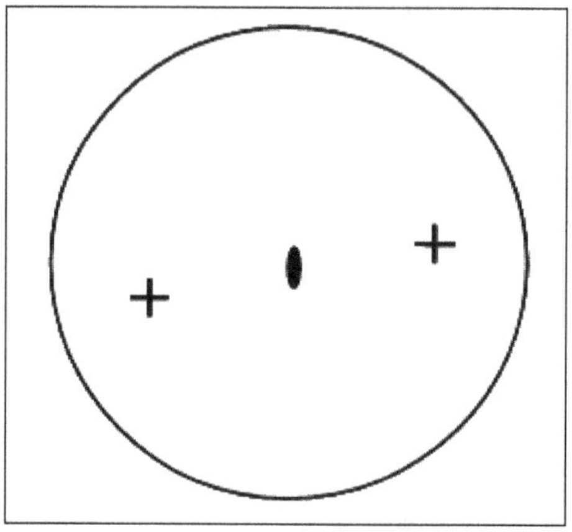

Figure shows Stereographic projection of 2-fold rotation axis. The cross in the upper, right side of the drawing is assigned the coodinates (x, y, z), the identity operation. If the 2-fold symmetry element is parallel with the b axis then the cross in the lower, left side of the drawing is assigned the coordinates (-x, y, -z).

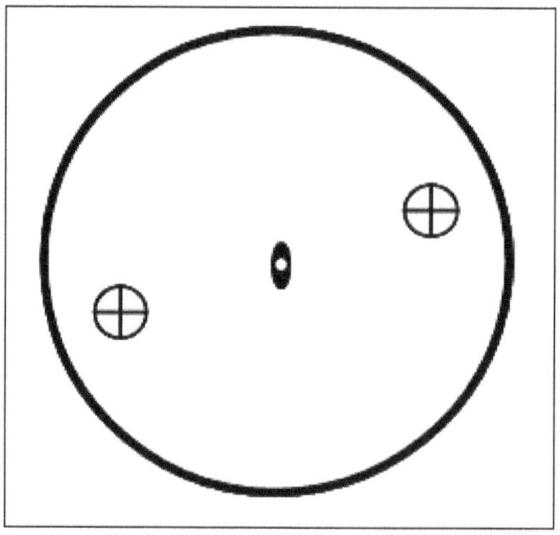

Figure: Stereographic projection of 2/m rotation.

This projection shows a circle in the lower left region with the coordinates (-x, -y, -z) that is a spot generated by the inversion center. That the combination of the 2-fold rotation axis and the inversion center leads to another operation--a mirror plane normal to the 2-fold. The relative coordinates of a point related by the mirror in b are (x, -y, z). The generation of a mirror by adding an inversion center to a 2-fold axis is an example of the closure property of the group. The mirror is shown by the strong outer circle of the projection.

The other point groups and their symmetry-related coordinates can be derived in a similar manner to that shown above. All 32 crystallographic point groups are shown in the stereographic projections below.

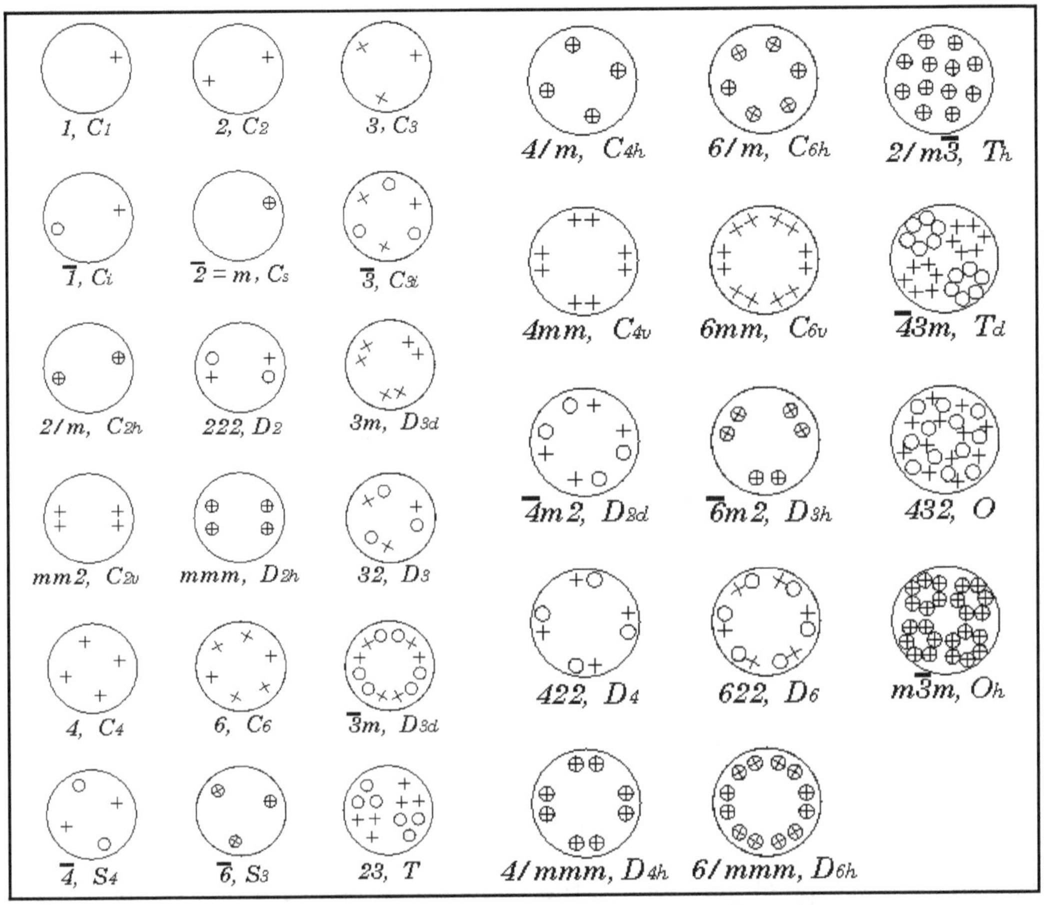

Figure: Stereographic projections of the 32 crystallographic point groups.

Micro-Translation

Rotational symmetry operations can be combined with translations of part of the unit cell creating entirely new symmetry operations. Proper rotations combined with translations give rise to operations described as *screw* axes. Mirror planes that are combined with translations give rise to *glide* plane operations.

The symbol for a screw axis is n_m where n indicates the type of rotation and the translation is (m/n) of the unit cell. A 2_1 operation is a rotation followed by a translation of 1/2 of the cell parallel with the rotation axis. There are a total of 11 types of screw axes, 2_1, 2_1, 3_1, 3_2, 4_1, 4_2, 4_3, 6_1, 6_2, 6_3, 6_4, and 6_5. Some pairs of screw axes rotate the objects in opposite directions. The opposite pairs are the 3_1 and 3_2, the 4_1 and 4_3, the 6_1 and 6_5, and the 6_2 and 6_4.

Glide operations occur when a mirror operation is followed by a translation. The translation directions either are parallel with a unit cell direction or are parallel with a combination of cell directions. Glide planes are described by the direction of the translation. The direction of the mirror plane is either explicitly given or is implied for example in the space group symbol. Glide plane operations exist in all three directions and in pairs of directions. Glides that translate by half of the cell in two different directions are called *n* glides. An object undergoes an *n* glide operation reflecting in the (001) plane would be translated by $(a + b)/2$.

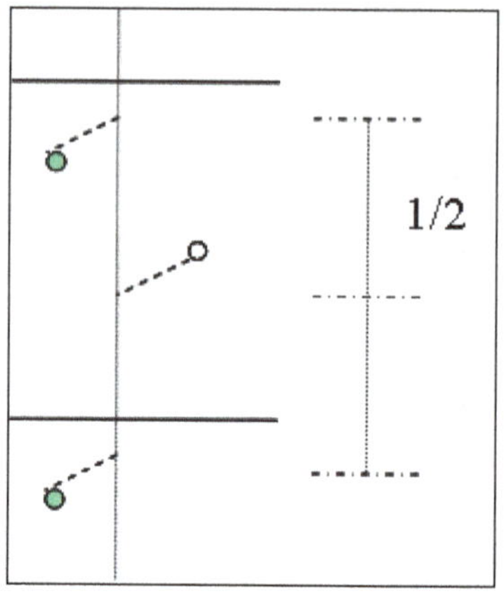

An example of an a, b, or c glide plane.

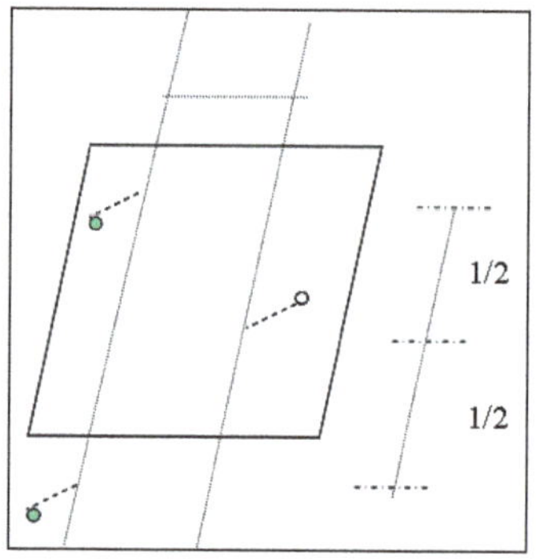

An example of an n glide plane.

There is one additional type of glide plane, the diamond glide, *d*. It occurs only in space groups with face- or body-centered cells, and is characterized by a translation of $(\pm a \pm b)/4$, $(\pm b \pm c)/4$, $(\pm c \pm a)/4$, or similar translations. As the denominator implies, 4 consecutive *d* glides are required to return an object to a lattice-translated version of itself.

Symmetry of the Reciprocal Lattice

The real cell parameters are determined by the relative positions of the reciprocal lattice points. Actually the reciprocal cell parameters are determined during a process known as indexing the diffraction pattern. From the reciprocal cell parameters the real cell parameters are then calculated according to the relations below:

- $a = b^* \times c^* = (b^* c^* \sin \alpha^*) / V^*$.
- $b = c^* \times a^* = (c^* a^* \sin \beta^*) / V^*$.
- $c = a^* \times b^* = (a^* b^* \sin \gamma^*) / V^*$.
- $V^* = 1/V = a^*b^*c^* (1\text{-}\cos^2 \alpha^* \text{-} \cos^2 \beta^* \text{-} \cos^2 \gamma^* + 2 \cos \alpha^* \cos \beta^* \cos \gamma^*)$.
- $\cos \alpha = (\cos \beta^* \cos \gamma^* - \cos \alpha) / (\sin \beta^* \sin \gamma^*)$.

- $\cos \beta = (\cos \gamma^* \cos \alpha^* - \cos \beta) / (\sin \gamma^* \sin \alpha^*)$.
- $\cos \gamma = (\cos \alpha^* \cos \beta^* - \cos \gamma) / (\sin \alpha^* \sin \beta^*)$.

Laue Class

The relative intensities in a diffraction pattern are dependent on the electron density distribution of the sample. This electron density distribution must follow the symmetry of the crystal itself. This symmetry is called the Laue class. The Laue class for a sample is described as one of the 11 centrosymmetric point groups. That the appropriate centrosymmetric point group or Laue class for a sample can be identified by adding a center of symmetry to the point group operations of the particular crystallographic point group of the sample.

Friedel's Law

The additional center of symmetry is due, at least approximately, to the fact that diffraction or interference effects are inherently centrosymmetric. The intensity of the $(\bar{h}\ \bar{k}\ \bar{l})$ point in the reciprocal lattice comes from scattering of the electron density that is parallel to the (h k l) planes in the crystal. Similarly, the intensity of the (h k l) point comes from the electron density in planes parallel with the $(\bar{h}\ \bar{k}\ \bar{l})$ planes in the crystal. But since the (h k l) planes and the $(\bar{h}\ \bar{k}\ \bar{l})$ planes are simply opposite in direction, then the intensities of the (h k l) and $(\bar{h}\ \bar{k}\ \bar{l})$ points in the reciprocal lattice should be, at least approximately the same. This equality relationship between the intensites of (h k l) and $(\bar{h}\ \bar{k}\ \bar{l})$ is called Friedel's law.

For the intensity data from a chiral compound, Friedel's law can be broken by the anomalous scattering of heavy atoms. In these data sets the reciprocal lattice has the same symmetry as the symmetry of the point group of the crystal. Thus if the point group symmetry of the crystal is shown to be 222 then the intensities would exhibit 222 symmetry. The anomalous scattering of heavy atoms is not a strong effect, so the intensities will still approximately exhibit Friedel's law.

Symmetry-Equivalent Intensities

One simple way to determine which data should have equivalent intensities based on symmetry is to consider the stereographic projection of the point group of the sample. From a simple examination of the stereographic projection, you can determine the symmetry-related (x, y, z) coordinates. Simply convert the x coordinates to h, the y coordinates to k, and the z coordinates to l values. The symmetry-equivalent intensities are then determined.

A monoclinic crystal, has the Laue symmetry of 2/m. The equivalent coordinates, assuming a b-unique axis, are given as (x, y, z), (-x, y, -z), (-x, -y, -z), and (x, -y, z). Thus the intensities of the (h k l), $(\bar{h}\ k\ \bar{l})$, $(\bar{h}\ \bar{k}\ \bar{l})$, and $(h\ \bar{k}\ l)$ lattice points should have equivalent values. That this also means that the intensities of the $(h\ k\ \bar{l})$, $(\bar{h}\ k\ l)$, $(\bar{h}\ \bar{k}\ l)$, and $(h\ \bar{k}\ \bar{l})$ should also be equivalent to each other but are not necessarily equivalent to (h k l), etc.

If a monoclinic compound is chiral then the intensities would be expected to have only 2 point group symmetry. Thus the intensities of the (h k l) data are equivalent to the intensities of the $(\bar{h}\ k\ \bar{l})$ data. Similarly, $I(h\ k\ \bar{l}) = I(\bar{h}\ k\ l)$; $I(h\ \bar{k}\ l) = I(\bar{h}\ \bar{k}\ \bar{l})$; and $I(h\ \bar{k}\ \bar{l}) = I(\bar{h}\ \bar{k}\ l)$.

If a crystal happens to have all three cell angles = 90.0° within experimental error then most workers would guess that the sample had orthorhombic symmetry. In most cases this guess would be correct, but not in all cases. If $I(h\,k\,l) = I(\bar{h}\,k\,\bar{l})$ and $I(h\,\bar{k}\,\bar{l}) = I(\bar{h}\,\bar{k}\,l)$, but $I(h\,k\,l) \neq I(h\,\bar{k}\,\bar{l})$, then the sample has monoclinic not orthorhombic Laue symmetry. The symmetry of the Laue class is dictated by the symmetry of the reciprocal lattice intensities not the apparent symmetry of the cell parameters. The cell parameters only dictate the possible highest symmetry of the sample.

Systematic Absences

Some symmetry operations can be readily identified by specific information in the intensities of the diffraction pattern. In particular, cell centering, screw axes, and glide plane operations can be identified by the fact that they cause certain groups of diffraction points to be systematically absent. These symmetry operations all include a micro-translation.

Consider a data set that has a c glide operation reflecting in the plane normal to the b axis. The symmetry operations could be (x, y, z) and (x, -y, z+1/2). If there are N atoms in the unit cell, then there are N/2 unique atoms. The summations below are over the j atoms and run from 1 to N/2:

- $\mathbf{F}(hkl) = \sum f_j \exp 2\pi i(hx_j + ky_j + lz_j)$.
- $\sum f_j \exp 2\pi i[hx_j - ky_j + l(1/2 + z_j)]$.

Consider the data with k = 0. For these data the structure factors become:

- $\mathbf{F}(h0l) = \sum f_j \exp 2\pi i(hx_j + lz_j)$.
- $\sum f_j \exp 2\pi i(hx_j + lzj) \exp 2\pi i(l/2)$.
- $\mathbf{F}(h0l) = \sum f_j \exp 2\pi i(hx_j + lz_j)\,[1 + \exp \pi il]$.

If l is an odd integer, then $\exp \pi il = -1$ and $\mathbf{F}(h0l) = 0$. If l is an even integer, then $\mathbf{F}(h0l)$ is probably not (but could accidentally be) 0. The systematic absence conditions for other symmetry operations can be derived in a similar manner as was done above.

The reflection condition indicates the data that may be present, other data with the same conditions would be systematically absent. The character n can be any integer. Thus the condition that for hkl, k + l = 2n+1, indicates that for the general class of peaks hkl that the sum of k + l should be positive for these peaks to have a measurable intensity, and that those peaks with k + l negative must not have a measurable intensity if the A centering symmetry is present in the symmetry of the data set.

It is always best to check for systematic absences, and hence for translation-containing symmetry operations, in the following order cell centering operations, then glide planes, and finally screw axes. This order is important because higher symmetry operations such as cell centering can mask lower symmetry operations such as glide planes and screw axes. Also, this order works from conditions that have the largest number of possible absences to the conditions with the least possible absences for a given sample.

That if a crystal exhibits the systematic absence condition for C cell centering then the fact that for okl, k = 2n does not necessarily indicate a b glide reflecting in a. This condition gives no new information, because it is a part of the cell centering condition.

Table: Systematically Absent Reflection Conditions.

Symmetry Element	Types	Reflection Condition
A centered	hkl	k + l = 2n
B centered		h + l = 2n
C centered		h + k = 2n
F centered		k + l = 2n, h + l = 2n, h + k = 2n
I centered		h + k + l = 2n
R (obverse)		-h + k + l = 3n
R (reverse)		h - k + l = 3n
Glide reflecting in **a**	okl	
b glide		k = 2n
c glide		l = 2n
n glide		k + l = 2n
d glide		k + l = 4n
Glide reflecting in **b**	hol	
a glide		h = 2n
c glide		l = 2n
n glide		h + l = 2n
d glide		h + l = 4n
Glide reflecting in **c**	hko	
b glide		k = 2n
a glide		h = 2n
n glide		k + h = 2n
d glide		k + h = 4n
Glide reflecting in (110)	hhl	
b glide		h = 2n
n glide		h + l = 2n
d glide		h + k + l = 4n
Screw ‖ [100]	hoo	
2_1, 4_2		h = 2n
4_1, 4_3		h = 4n
Screw ‖ [010]	oko	
2_1, 4_2		k = 2n
4_1, 4_3		k = 4n
Screw ‖ [001]	ool	
2_1, 4_2, 6_3		l = 2n
3_1, 3_2, 6_2, 6_4		l = 3n
4_1, 4_3		l = 4n

$6_1, 6_5$		l = 6n
Screw ǀǀ [110]	hho	
2_1		h = 2n

Space Groups

When the 7 crystal systems are combined with the 14 Bravais lattices, the 32 point groups, screw axes, and glide planes, Arthur Schönflies, Evgraph S. Federov, and H. Hilton were able to describe the 230 unique space groups. A space group is a group of symmetry operations that are combined to describe the symmetry of a region of 3-dimensional space, the unit cell. In point groups, the symmetry elements all pass through one point in the object. In space groups, symmetry elements need not intersect at a single point, although some operations may coincidentally intersect at various points in the cell.

Chiral compounds that are prepared as a single enantiomer can crystallize in only a subset of 65 space groups that do not have improper rotations such as centers of symmetry, mirror planes, glides, or $\bar{3}$, $\bar{4}$, or $\bar{6}$ axes.

A space group is designated by a capital letter identifying the lattice type (P, A, F, etc.) followed by the point group symbol in which the rotation and reflection elements are extended to include screw axes and glide planes. That the point group symmetry for a given space group can be determined by removing the cell centering symbol of the space group and replacing all screw axes by similar rotation axes and replacing all glide planes with mirror planes. The point group symmetry for a space group describes the true symmetry of its reciprocal lattice. The following table lists the unique space groups separated by crystal system and Laue class.

Table: Space groups based on crystal system and Laue class.

Crystal System	Laue Class	Space Group
triclinic	$\bar{1}$	$P1$, **$P\bar{1}$**
monoclinic	$2/m$	$P2$, $P2_1$, $C2$, Pm, Pc, Cm, Cc, **$P2/m$**, **$P2_1/m$**, **$C2/m$**, **$P2/c$**, **$P2_1/c$**, **$C2/c$**
orthorhombic	mmm	$P222$, $P222_1$, $P2_12_12$, $P2_12_12_1$, $C222_1$, $C222$, $F222$, $I222$, $I2_12_12_1$, $Pmm2$, $Pmc2_1$, $Pcc2$, $Pma2$, $Pca2_1$, $Pnc2$, $Pmn2_1$, $Pba2$, $Pna2_1$, $Pnn2$, $Cmm2$, $Cmc2_1$, $Ccc2$, $Amm2$, $Aem2$, $Ama2$, $Aea2$, $Fmm2$, $Fdd2$, $Imm2$, $Iba2$, $Ima2$, **$Pmmm$**, **$Pnnn$**, **$Pccm$**, **$Pban$**, **$Pmma$**, **$Pnna$**, **$Pmna$**, **$Pcca$**, **$Pbam$**, **$Pccn$**, **$Pbcm$**, **$Pnnm$**, **$Pmmn$**, **$Pbcn$**, **$Pbca$**, **$Pnma$**, **$Cmcm$**, **$Cmce$**, **$Cmmm$**, **$Cccm$**, **$Cmme$**, **$Ccce$**, **$Fmmm$**, **$Fddd$**, **$Immm$**, **$Ibam$**, **$Ibca$**, **$Imma$**
tetragonal	$4/m$	$P4$, $P4_1$, $P4_2$, $P4_3$, $I4$, $I4_1$, $P\bar{4}$, $I\bar{4}$, **$P4/m$**, **$P4_2/m$**, **$P4/n$**, **$P4_2/n$**, **$I4/m$**, **$I4_1/a$**
tetragonal	$4/mmm$	$P422$, $P42_12$, $P4_122$, $P4_12_12$, $P4_222$, $P4_22_12$, $P4_322$, $P4_32_12$, $I422$, $I4_122$, $P4mm$, $P4bm$, $P4_2cm$, $P4_2nm$, $P4cc$, $P4nc$, $P4_2mc$, $P4_2bc$, $I4mm$, $I4cm$, $I4_1md$, $I4_1cd$, $P\bar{4}2m$, $P\bar{4}2c$, $P\bar{4}2_1m$, $P\bar{4}2_1c$, $P\bar{4}m2$, $P\bar{4}c2$, $P\bar{4}b2$, $P\bar{4}n2$, $I\bar{4}m2$, $I\bar{4}c2$, $I\bar{4}2m$, $I\bar{4}2d$, **$P4/mmm$**, **$P4/mcc$**, **$P4/nbm$**, **$P4/nnc$**, **$P4/mbm$**, **$P4/mnc$**, **$P4/nmm$**, **$P4/mcc$**, **$P4_2/mmc$**, **$P4_2/mcm$**, **$P4_2/nbc$**, **$P4_2/nnm$**, **$P4_2/mbc$**, **$P4_2/mnm$**, **$P4_2/nmc$**, **$P4_2/ncm$**, **$I4/mmm$**, **$I4/mcm$**, **$I4/amd$**, **$I4/_1acd$**
trigonal	$\bar{3}$	$P3$, $P3_1$, $P3_2$, $R3$, **$P\bar{3}$**, **$R\bar{3}$**.
trigonal	$3m$	$P312$, $P321$, $P3_112$, $P3_121$, $P3_212$, $P3_221$, $R32$, $P3m1$, $P31m$, $P3c1$, $P31c$, $R3m$, $R3c$, **$P\bar{3}1m$**, **$P\bar{3}1c$**, **$P\bar{3}m1$**, **$P\bar{3}c1$**, **$R\bar{3}m$**, **$R\bar{3}c$**
hexgonal	$6/m$	$P6$, $P6_1$, $P6_5$, $P6_2$, $P6_4$, $P6_3$, $P\bar{6}$, **$P6/m$**, **$P6_3/m$**
hexagonal	$6/mmm$	$P622$, $P6_122$, $P6_522$, $P6_222$, $P6_422$, $P6_322$, $P6mm$, $P6cc$, $P6_3cm$, $P6_3mc$, $P\bar{6}m2$, $P\bar{6}c2$, $P\bar{6}2m$, $P\bar{6}2c$, **$P6/mmm$**, **$P6/mcc$**, **$P6_3/mcm$**, **$P6_3/mmc$**
cubic	$m\bar{3}$	$P23$, $F23$, $I23$, $P2_13$, $I2_13$, **$Pm\bar{3}$**, **$Pn\bar{3}$**, **$Fm\bar{3}$**, **$Fd\bar{3}$**, **$Im\bar{3}$**, **$Pa\bar{3}$**, **$Ia\bar{3}$**
cubic	$m\bar{3}m$	$P432$, $P4_232$, $F432$, $F4_132$, $I432$, $P4_332$, $P4_132$, $I4_132$, $P\bar{4}3m$, $F\bar{4}3m$, $I\bar{4}3m$, $P\bar{4}3n$, $F\bar{4}3c$, $I\bar{4}3d$, **$Pm\bar{3}m$**, **$Pn\bar{3}n$**, **$Pm\bar{3}n$**, **$Pn\bar{3}m$**, **$Fm\bar{3}m$**, **$Fm\bar{3}c$**, **$Fd\bar{3}m$**, **$Fd\bar{3}c$**, **$Im\bar{3}m$**, **$Ia\bar{3}d$**

The space groups in bold are centrosymmetric. The mathematically-unique space groups. Some non-standard space groups are reported in a publication so that the cell parameters of several related compounds can be more easily compared. For example, if the main compound in a paper is found to be in the monoclinic space group *C*2/*c*, but a related compound is found to be triclinic, some authors would determine the structure of this second compound in the non-standard space group $C\bar{1}$ if the cell parameters in $C\bar{1}$ were similar to the cell parameters of the main compound. Finally some authors would publish structures in non-standard space groups because the basis vectors of the related cells would produce cell angles that were closer to 90 °. For example, structures are often published in *Pn* rather than *Pc* or in *I*2/*a* rather than *C*2/*c* because the β angle would be much closer to 90°. When β angles are closer to 90 °, the refinement of the x and z coordinates of the atoms show much less correlation, and converge to form chemically-reasonable structures more quickly.

The following tables list the graphical and typed symbols used to describe symmetry operations in the International Tables for Crystallography, Vol A. The symbol ⊥ is used to indicate that the operation is perpendicular (normal) to the page.

Table: Symmetry planes normal to the plane of projection.

Symmetry plane	Graphical symbol	Translation	Symbol
Reflection plane		None	*m*
Glide plane		1/2 along line	*a*, *b*, or *c*
Glide plane		1/2 normal to plane	*a*, *b*, or *c*
Double glide plane		1/2 along line & 1/2 normal to plane	*e*
Diagonal glide plane		1/2 along line & 1/2 normal to plane	*n*
Diamond glide plane		1/4 along line & 1/4 normal to plane	*d*

Table: Symmetry planes parallel to the plane of projection.

Symmetry plane	Graphical symbol	Translation	Symbol
Reflection plane		None	*m*
Glide plane		1/2 along arrow	*a*, *b*, or *c*
Double glide plane		1/2 along either arrow	*e*
Diagonal glide plane		1/2 along the arrow	*n*
Diamond glide plane	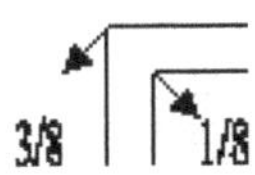	1/8 or 3/8 along the arrows	*d*

Table: Symmetry Axes.

Symmetry Element	Graphical Symbol	Translation	Symbol
Identity	None	None	1
2-fold ⊥ page		None	2

2-fold in page		None	2
2 sub 1 ⊥ page		1/2	2_1
2 sub 1 in page		1/2	2_1
3-fold		None	3
3 sub 1		1/3	3_1
3 sub 2	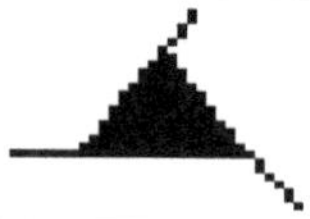	2/3	3_2
4-fold		None	4
4 sub 1		1/4	4_1
4 sub 2		1/2	4_2
4 sub 3		3/4	4_3
6-fold		None	6
6 sub 1		1/6	6_1
6 sub 2		1/3	6_2
6 sub 3		1/2	6_3
6 sub 4	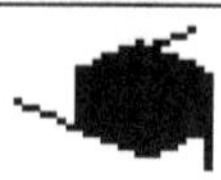	2/3	6_4
6 sub 5		5/6	6_5
Inversion	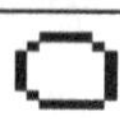	None	$\bar{1}$
3 bar		None	$\bar{3}$

4 bar		None	$\bar{4}$
6 bar		None	$\bar{6} = 3/m$
2-fold and inversion		None	2/m
2 sub 1 and inversion		None	$2_1/m$
4-fold and inversion		None	4/m
4 sub 2 and inversion		None	$4_2/m$
6-fold and inversion		None	6/m
6 sub 3 and inversion		None	$6_3/m$

Determining the Space Group of a Material

Identifying the proper space group for a given sample and its diffraction pattern is often straight forward. The Laue class is determined first. Then the possible systematic absences are determined to identify the appropriate cell centering conditions, glide planes, and screw axes, if any. From the Laue class and the symmetry operations identified by the systematic absences, the choice of space group(s) is usually limited one unique or to a very few.

If the space group is not uniquely determined then the structure solution and refinement steps are tried with the different possible space groups until the structure determination is completed. Try the space groups beginning with the highest symmetry first. If the chosen space group is not the highest symmetry space group, then be sure to check for additional symmetry elements in the structure using either the PLATON program or the check if program.

Additional information about the compound is often used to reduce the choice of space group to a single one. If the compound is chiral and is expressed as a single enantiomorphism, for example, then the compound can only crystallize in one of the 65 chiral space groups.

Merging Data to Determine Laue Class

The Laue class is usually found by calculating the merging Rint of the symmetry-equivalent data for the given Laue class. An appropriate Rint for identifying the Laue class would be a value & 0.06. Values larger than this usually suggests that the sample crystallized in a lower symmetry Laue group. Symmetry-equivalent intensity data are merged using the following relationship:

$$F^2 = \sum \omega_j F_j^2 / \sum \omega_j$$

where the summations are over the set of symmetry-equivalent data. Sometimes outliers are either removed from the merged set or are given lower weights to reduce their effects on the subsequent refinement:

$$Rint = \sum [\sum |F_j^2 - \langle F^2 \rangle|] / \sum [(\sum F_j^2)/n]$$

Where the inner sums are over the symmetry-equivalent reflections and the outer sums are over the unique hkl data. The term n is the number of equivalent data for a given hkl being merged.

Tests for a Center of Symmetry

If the compound is not a single enantiomorphism, then statistical tests for a center of symmetry can be performed. Probability distributions for centric and acentric unit cells have been derived22 and are given below. These derivations assume that the electron density is randomly and uniformly distributed throughout the unit cell:

- $P_{-1}(|F|) = [(2)^{1/2} / (\pi S)^{1/2}] \exp(-|F|^2 / 2 S)$ (centric).
- $P_1(|F|) = (2 |F| / S) \exp(-|F|^2 / S)$ (acentric).

where,

$$S = \sum f_j^2$$

S is a function of the scattering factors for the atoms and thus will decrease for increasing sin (θ / λ). To simplify this functional dependence of the distributions on θ, most statistical tests utilize normalized structure factors, Ehkl.

$$E_{hkl}^2 = F_{hkl}^2 / (\varepsilon \sum f_j^2).$$

where $f_j = f_j o \exp(-B \sin^2\theta/\lambda 2)$ is the scattering factor for the jth atom and ε is an integer, 1 or greater, that corrects for the fact that some classes of reflections have expectation values that are less than $\sum f_j^2$ by an integer amount.

Since $\langle |Eh|^2 \rangle = 1 = S$, the distributions become:

- $P_{-1}(|E|) = [(2 / \pi)^{1/2}] \exp(-|E|^2 / 2)$ (centric).
- $P_1(|E|) = 2 |E| \exp(-|E|^2)$ (acentric).

These two distributions are completely independent of the variation of θ and are significantly different from each other. The differences in the two distributions indicate that centric unit cells should have more reflections with very strong and very weak values compared to acentric unit cells. Acentric unit cells typically tend to have more even distributions of data with significantly fewer weak data than centric cells. Comparing the distribution of a measured data set with the two theoretical distributions could establish the presence or absence of a center of symmetry in the crystal under consideration.

Mean values for a variety of functions of |E| may be estimated from the two theoretical distributions. These values are presented in the following Table. Most computer programs that test for centric unit cells calculate the mean values for these functions of E and compare these experimental values with the theoretical values.

Table: Theoretical Values for Selected Functions of |E|.

	Centric	Acentric
$<\|E\|^2>$	1.000	1.000
$<\|E^2-1\|>$	0.968	0.736
$<\|E\|>$	0.798	0.886

Three caveats must be remembered when using these tests. First, the intensity data must be relatively strong. Weak data sets, having $F^2/\sigma < 6.0$, have too many weak data to give reliable results for this test. Second, these theoretical values were derived assuming a uniform distribution of electron density in the unit cell. The presence of heavy atoms or of an unusual distribution of atoms tends to make the experimental values assume a centric distribution. When the tests indicate an acentric unit cell, the tests are likely to be correct. When the tests suggest a centric unit cell, the tests may be correct or may simply indicate a skewed distribution of electron density in the unit cell. Finally, if the tests indicate a hyper acentric distribution, e.g. $<|E^2-1|> < 0.6$, then it is likely that the sample is twinned.

Polar Space Groups

Some space groups do not contain symmetry operators that fix the origin of the unit cell in one or more directions. These space groups can be identified by the examining the symmetry-equivalent positions for the space group. For any of the three coordinate directions there are no symmetry operations that invert that coordinate (x -> -x), (y -> -y), or (z -> -z). For these polar space groups, the origin must be fixed by the atoms. This may be accomplished either by holding the coordinate(s) in the polar direction(s) fixed for one atom during refinement or by applying a constraint to hold some function, usually the mean value, of all coordinates in the given direction(s) fixed. If the polar axis is established by fixing the coordinate or coordinates of one atom, then selecting the heaviest atom in the structure for this purpose will lead to a more stable refinement.

Inversion of a Crystal Structure

Very few methods of solving a structure are able to properly pick the correct *absolute configuration* of a crystal structure. For these samples, the structure is roughly determined and refined. The configuration is then checked either by means of anomalous scattering or by knowing and checking the configuration of the compound itself. For example, naturally occuring peptides should have most helices with a right-handed twist. There may be a few short helices with a left-handed twist, but the nearly all should have a right-handed twist.

If the structure is determined to have the incorrect handedness, then the structure is simply inverted through some symmetry point such as (0, 0, 0) or (0.5, 0.5, 0.5) and the refinement is completed. The space group must also be changed if the structure crystallizes in one of the 11 enantiomorphous pairs of space groups. If the sample crystallizes in one of 7 special space groups, then inversion must be through a special point.

The 230 space groups include 11 space groups that occur as enantiomorphous pairs: $P3_1$, $P3_2$; $P3_112$, $P3_212$; $P3_121$, $P3_221$;$P4_1$, $P4_3$; $P4_122$, $P4_322$; $P4_12_12$, $P4_32_12$; $P6_1$, $P6_5$; $P6_2$, $P6_4$; $P6_122$, $P6_522$; $P6_222$, $P6_422$; and $P4_132$, $P4_332$. When a sample crystallizes in one of these space groups, one of the pair is arbitrarily chosen for the structure solution and at least partial refinement. If it is later determined that the wrong space group of the pair was chosen initially, then the atoms are inverted and the alternate space group of the pair is used in further refinements.

For 7 other space groups' inversion must be carried out through some point in the structure other than (0, 0, 0) or (0.5, 0.5, 0.5). A table listing these seven space groups and appropriate inversion points is listed below.

Table: Space Groups with Non-origin Inversion Points.

Space Group	Inversion Point	SHELXTL MOVE
Fdd2	0.125 0.125 0.500	MOVE 0.25 0.25 1 -1
$I4_1$	0.500 0.250 0.500	MOVE 1 0.5 1 -1
$I4_122$	0.500 0.250 0.125	MOVE 1 0.5 0.25 -1
$I4_1md$	0.500 0.250 0.500	MOVE 1 0.5 1 -1
$I4_1cd$	0.500 0.250 0.500	MOVE 1 0.5 1 -1
$I\bar{4}2d$	0.500 0.250 0.125	MOVE 1 0.5 0.25 -1
$F4_132$	0.125 0.125 0.125	MOVE 0.25 0.25 0.25 -1

Special Positions

Many space groups contain simple symmetry operations, such as inversion centers, rotation axes, and mirror planes. Usually the locations of these symmetry elements are either fixed by convention, or fixed relative to other symmetry operations in the cell. These special positions are located on points (centers of symmetry), lines (rotations axes) or planes (mirror planes). Also, there are combinations of symmetry operations such as 2/m points or mm lines.

Because atoms on some of these special positions are a constant distance from their symmetry-related atoms, these special positions can give rise to systematically absent intensity data. These systematic absences can be derived in a similar manner as the systematic absence conditions for glide planes and screw axes derived above.

That these systematic absence conditions apply only to atoms that sit on the respective special positions. This can be a problem when the atom(s) on a special position are very heavy and the remaining atoms are in general positions are much lighter and are, in general, evenly distributed in the cell. Under these circumstances, the systematic absence due to the atom on the special position could be confused as a general systematic absence, obscuring the space group symmetry. Thus when a structure cannot be solved, and other tricks have been tried, it may be necessary to lower the symmetry to try to solve the structure.

Space Group Names

Although the Hermann-Mauguin names for space groups will always be used in publications and in the cifs accompanying the publications, another set of names for space groups is also included in cif files. This alternate description of space groups was developed by Sid Hall to give a unique name to each space group and space group setting. However, all of the ambiguities in H-M space group names are resolved by looking at either the cell parameters or the symmetry operators.

Crystal Systems

Crystal system is a method of classifying crystalline substances on the basis of their unit cell. There are seven unique crystal systems. The simplest and most symmetric, the cubic (or isometric) system, has the symmetry of a cube. The other six systems, in order of decreasing symmetry, are hexagonal, tetragonal, rhombohedral (also known as trigonal), orthorhombic, monoclinic and triclinic. Five Symmetry classes fall in the Isometric System by virtue of their symmetry. The normal class is known as galena type.

Axes of Symmetry

- 13 in all,
- 3 are axes of four-fold symmetry,
- 4 are axes of three-fold symmetry,
- 6 are axes of two-fold symmetry,
- The three axes of four-fold symmetry are chosen as the crystallographic axes.

Planes of Symmetry

- 9 in all,
- 3 planes of symmetry are at right angles to each other and are termed the principal (axial) planes,
- 6 planes of symmetry are diagonal in position and bisect the angles between the principal planes,
- It has a Centre of symmetry.

Following are the forms that commonly develop in the crystals belonging to Isometric System:

- Cube: A form bounded by six similar square faces, each of which is parallel to two of three crystallographic axes and meets the third axis.

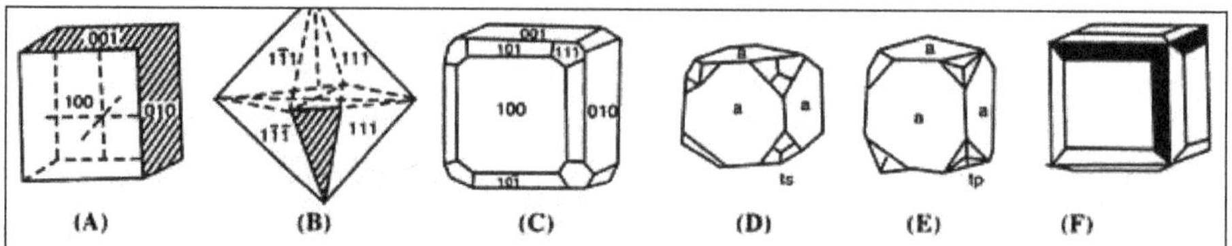

(A) Cube, (B) Octahedron and (C) Dodecahedron (D) Trisoctahedron (E) Trapezohedron (F) Tetrehexahedron

- Octahedron: A form bounded by eight similar faces, each of the shape of an equilateral triangle, each meeting the three crystallographic axes at equal distances.
- Dodecahedron: It is a form with twelve similar faces each of which is parallel to one of the three crystallographic axes and meets the other two at equal distances.
- Trisoctahedron (hhl): A form of twenty four (24) faces; each face meeting two axes at unit length and to the third at greater than unity. Faces occur in eight groups of three each.
- Trapezohedron (hll): A form of twenty four (24) faces each face meeting one axes at unit length and to the other two at greater than unity. Each face is a trapezium.
- Tetra-Hexahedron (hol): Twenty four (24) faces; each face is parallel to one axis and meets other two at unequal distance which are simple multiple of each other; faces occur in six groups of four each.
- Hexaocatahedron (hkl): Forty eight (48) faces; each face meets the three axis at unequal distances.
- Other Classes: Isometric system comprises five symmetry classes in all.

Beside the normal class, following three classes are also represented among the minerals:

Pyritohedron Class (Pyrite Type)

- Symmetry: 7 Axes of symmetry, of which, 3 are axial axes of two-fold symmetry, and 4 are diagonal axes of two fold symmetry. 3 Planes of symmetry. A Centre of symmetry.
- Forms: Pyritohedron and Diploid are two typical forms of this symmetry class. Pyritohedron is a hemihedral form having twelve faces and a general symbol (hko).

Diploid is a closed form of twenty-four faces that typically occur in pairs (hence the name) and have a symbol (hki).

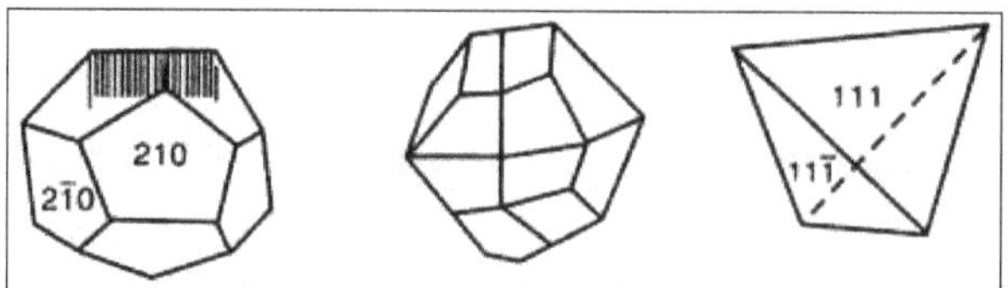

A – Pyritohedron, B – Diploid, C – Tetrahedron

Tetrahedral Class (Tetrahedrite Type)

- Symmetry: 7 axes of symmetry (as in pyrite type), 6 planes of symmetry, diagonal, no center of symmetry.
- Forms: Most typical form of this class is a four sided solid in which each face is an equilateral triangle. It is termed tetrahedron. It has a general symbol of (111).

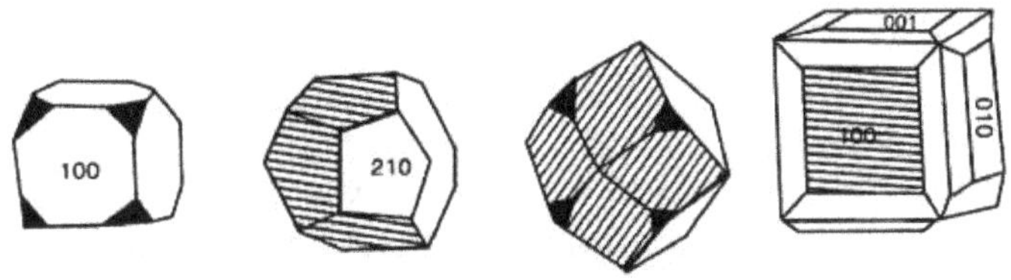

Some Isomeric Minerals A – Galena, B – Pyrite, C – Magnetite, D – Fluorite

Plagiohedral Class (Cuprite Type)

- Symmetry: 13 axis of symmetry (as in normal class) No planes of symmetry. No Centre of symmetry.
- Forms: Icositedraherons, each of 24 faces, with a symbol (hkl) and commonly enantiomorphous in character are typical forms of this class.

Examples of Isometric Minerals: A vast number of common minerals crystallize in isometric system. Following are few examples:

- Galena,
- Pyrite,
- Sphalerite,
- Spinel,
- Magnetite,
- Sodalite,
- Garnet,
- Diamond,
- Flourite,
- Leucite.

Tetragonal System

The tetragonal system includes all those crystals which can be referred to three crystallographic axes so that:

- Two axes are equal, interchangeable and horizontal in position.
- The third axis is either longer or shorter than the other two and is vertical in position.
- All the three axes are mutually at right angles to each other.

Axial Diagram

In study position, the crystal is so held that the longer or shorter axis is vertical in position. The one of the horizontal axes is made to run parallel to the observer and the other form front backward.

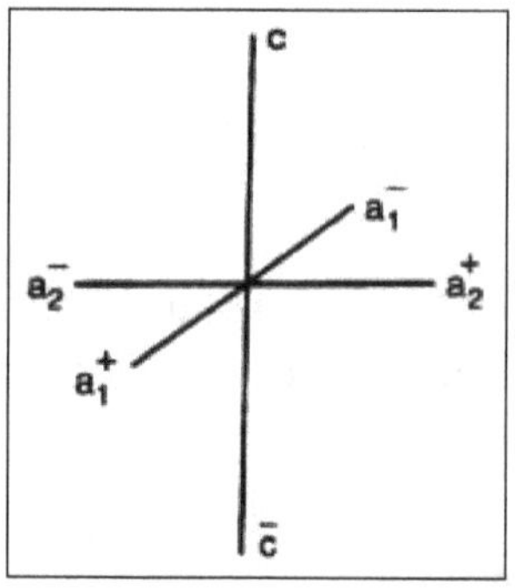

The horizontal axis running away from the observer is designated as a_1, that running from right to left is a_2, and the vertical axis is designated as c axis. Axial ratio for the crystal of this system is expressed as a: c.

Classes

The Tetragonal system includes seven symmetry classes.

Normal Class (Zircon Type)

The symmetry of the normal class (Zircon type) is as follows:

- Axis of Symmetry: 5 in all; of these, 2 axes are horizontal and of two fold symmetry, 2 axes are horizontal bisecting the angles made by first set of horizontal axis; they are also of two fold symmetry. 1 axis is vertical and of four folds symmetry (hence tetragonal forms).
- Planes of symmetry: 5 in all; of these, 1 plane is horizontal, 2 planes are vertical, 2 planes are vertical diagonal. There is a centre of symmetry.

Forms

Following forms are met with in the crystals of the Tetragonal System:

- Basal Pinacoid (001): An open form with two similar faces. Each face is parallel to the two horizontal crystallographic axes and meets the third, vertical axis at unit length.

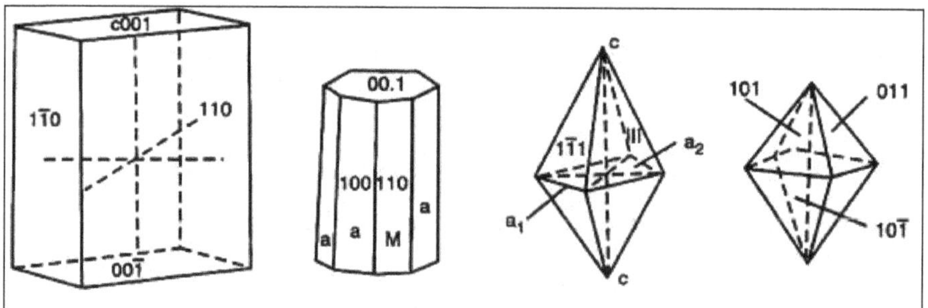

Forms in Tetragonal System

- Prisms: These are open forms of four or eight faces in which each face is essentially parallel to the vertical crystallographic axes.

Depending on their relation with the other two axes, three types of prisms are recognized in the tetragonal system:

- Prism of First Order (110): Four faces each face meeting the two horizontal axes – at unit lengths and being parallel to the vertical axis.
- Prism of Second Order (100): Four faces; each face intercepts only one of the horizontal axes and is parallel to the second axis besides being parallel to the vertical axis.
- Prism of Third Order or Ditetragonal Prism (hko): Eight faces each face intercepting the two horizontal axes at different lengths.

Pyramids

These are closed forms of eight or sixteen faces in which each face essentially meets the vertical crystallographic axis. Three types of pyramids are recognized on the basis of their relationship with other crystallographic axis:

- Pyramid of First Order (hhl): Eight faces; each face cuts the two horizontal axis at equal length besides intersecting the vertical axis.
- Pyramid of Second Order (hol): Eight faces; each face cuts only one of the two horizontal axes; it is parallel with one horizontal crystallographic axis.
- Pyramid of Third Order (hkl): Also called ditetragonal pyramid is a closed form of 16 faces in which each face cuts all the three crystallographic axes at unequal lengths.

Other Classes: Of the remaining six classes, the following three are well represented among minerals:

- Tripyramidal class,
- Pyramidal Hemimorphic class,
- Sphenoidal class.

Examples of Tetragonal Minerals:

- Zircon,
- Vesuvianite,
- Rutile,
- Wernerite,
- Cassiterite,
- Octahedrite,
- Chalcopyrite,
- Scheelite,

- Wulfenite,
- Braunite.

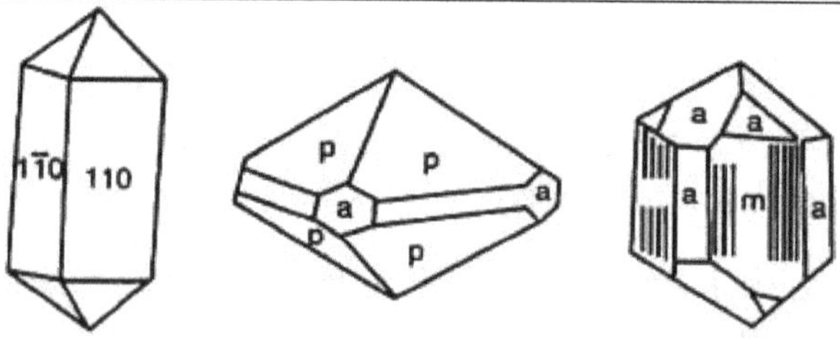

Some Minerals of Tetragonal System A – Zircon, B – Vesuvianite, C – Rutile

Crystal Lattices

A three-dimensional lattice can be visualized best as an infinite periodic array of points, which are the termini of the vectors:

$$\mathbf{l}_{uvw} = u\mathbf{a} + v\mathbf{b} + w\mathbf{c}, \quad u, v, w \text{ all integers.}$$

The parallelepiped determined by the basis vectors **a**, **b**, **c** is called a (primitive) *unit cell* of the lattice, **a**, **b** and **c** are a primitive *basis of the lattice*. The number of possible lattice bases is infinite.

For the investigation of the properties of lattices, appropriate bases are required. In order to select suitable bases, transformations may be necessary.

Topological Properties of Lattices

The treatment of the topological properties is restricted here to the consideration of the neighbourhood of a lattice point. For this purpose, the *domain of influence* is introduced. The domain of a particular lattice point consists of all points in space that are closer to this lattice point than to any other lattice point or at most equidistant to it. To construct the domain, the selected lattice point is connected to all other lattice points. The set of planes perpendicular to these connecting lines and passing through their midpoints contains the boundary planes of the domain of influence, which is thus a convex polyhedron. (Niggli and Delaunay used the term 'domain of influence' for the interior of the convex polyhedron only). Without the use of metrical properties, Minkowski proved that the maximal number of boundary planes resulting from this construction is equal to $2(2^n - 1)$, where n is the dimension of the space. The minimal number of boundary planes is 2n. Each face of the polyhedron represents a lattice vector. Thus, the topological, metrical and symmetry properties of infinite lattices can be discussed with the aid of a single finite polyhedron, namely the domain of influence.

Special Bases to Lattices

Different procedures are in use to select special bases of lattices. The reduction procedures employ metrical properties to develop a sequence of basis transformations which lead to a *reduced basis and reduced cell.*

Another possibility is to make use of the symmetry properties of lattices. This procedure, with the additional aid of standardization rules, leads to the *conventional crystallographic basis* and cell. In addition to translational symmetry, a lattice possesses point-group symmetry. No crystal can have higher point-group symmetry than the point group of its lattice, which is called its *holohedry* the seven holohedries in three dimensions and the four in two dimensions form the basis for the classification of lattices. It may be shown by an algebraic approach or a topological one that the arrangement of the symmetry elements with respect to the lattice vectors is not arbitrary but well determined. Taking as basis vectors lattice vectors along important symmetry directions and choosing the origin in a lattice point simplifies the description of the lattice symmetry operations. Note that such a basis is not necessarily a (primitive) basis of the lattice. The choice of a basis controlled by symmetry is not always unique; in the monoclinic system, for example, one vector can be taken parallel to the symmetry direction but the other two vectors, perpendicular to it, are not uniquely determined by symmetry.

Table: Lattice point-group symmetries - Two dimensions.

Lattice point group	2	$2mm$	$4mm$	$6mm$
Crystal family	m	o	t	h
	monoclinic (oblique)	orthorhombic (rectangular)	tetragonal (square)	hexagonal

Table: Lattice point-group symmetries - Three dimensions.

Lattice point group	$C_i \equiv \bar{1}$	$C_{2h} \equiv 2/m$	$D_{2h} \equiv mmm$	$D_{4h} \equiv 4/mmm$	$D_{3d} \equiv \bar{3}m$	$D_{6h} \equiv 6/mmm$	$O_h \equiv m\bar{3}m$
Crystal family	a	m	o	t	h		c
	anorthic (triclinic)	monoclinic	orthorhombic	tetragonal	hexagonal		cubic

The choice of conventions for standardizing the setting of a lattice depends on the purpose for which it is used. The several sets of conventions rest on two conflicting principles: symmetry considerations and metric considerations. The following rules to defining a *conventional basis* is taken from Donnay; they deal with the conventions based on symmetry:

- Each basis vector is a lattice vector from the origin to the nearest node on the related row. The basis must define a right-handed coordinate system.
- The basis vectors for a cubic lattice are parallel to the fourfold axes.
- In a *hexagonal* lattice, the basis vector parallel to the sixfold axis is labelled c. The remaining two basis vectors are taken along equivalent twofold axes and they must include an angle of 120; from the two possible sets, the shorter vectors are chosen.
- For *rhombohedral* lattices, two descriptions are given in the present edition, as in earlier ones. The first description which results in the conventional cell uses 'hexagonal axes'. In this case, c is taken along the threefold axis. The remaining two vectors are chosen along equivalent twofold axes, which include an angle of 120; they are oriented so that lattice points occur at 2/3; 1/3; 1/3 and 1/3; 2/3; 2/3 (obverse setting). The reverse setting ðo;

0; 0; 1/3; 2/3; 1/3; 2/3; 1/3; 2/3Þ is not used in the space-group tables. The second description uses 'rhombohedral axes': **a, b** and **c** are the shortest three non-coplanar lattice vectors symmetry-equivalent with respect to the threefold axis.

- In a *tetragonal* lattice, the vector **c** is along the fourfold axis, and **a** and **b** are chosen along equivalent twofold axes perpendicular to each other. From the two possible sets, the shorter vectors are chosen.
- In an *orthorhombic* lattice, **a, b** and **c** must be taken along the three twofold axes.
- For *monoclinic* lattices, two 'settings' are given in the present edition. In one setting, the only symmetry direction is labelled **b** (b-unique setting). The basis vectors **a** and **c** are chosen to be the shortest two vectors in the net plane perpendicular to **b**, the angle should be non-acute.

  This occurs if,

  $0 \leq -2\mathbf{a} \cdot \mathbf{c} \leq \min(a^2, c^2)$

  In the other setting, the symmetry direction is labelled **c**. In this case, **a** and **b** are the shortest two vectors in the net plane perpendicular to **c** and the angle should be non-acute. The b-unique setting is considered to be the standard setting.
- The reduced basis is used to describe a *triclinic* (= *anorthic*) lattice.

The metric parameters of the conventional basis are called *lattice parameters*. For the purpose of identification, additional metric rules are to be applied to make the labelling unique; they can be found in the introduction to *Crystal Data*.

When the above rules have been applied, it may occur that not all lattice points can be described by integral coordinates. In such cases, the unit cell contains two, three or four lattice points. The additional points may be regarded as *centrings* of the conventional cell. They have simple rational coordinates. For a conventional basis, the number of lattice points per cell is 1, 2, 3 or 4.

In two dimensions, only two centring types are needed:

p: no centring (primitive);

c: face centred.

In three dimensions, the following centring types are used:

P: no centring (primitive);

I: body centred (innenzentriert);

F: all-face centred;

A; B; C: one-face centred, either (**b, c**) or (**c, a**) or (**a, b**);

R: hexagonal cell rhombohedrally centred.

In orthorhombic and monoclinic lattices, some differently centred cells can be transformed into each other without violating the symmetry conditions for the choice of the basis vectors. In these cases, the different centred cells belong to the same *centring mode*. In the orthorhombic case, the three types of one-face-centred cells belong to the same centring mode because the symbol of the cell depends only on the labelling of the basis vectors; C is usually preferred to A and B as the standard setting; the centring mode is designated S (*seitenfla¨chenzentriert*). In the monoclinic case (b-unique setting), A, I and C may occur if the basis is chosen. They can be transformed into each other without changing the symmetry direction. C is used for the standard setting; it represents the centring mode S. The vectors **a**, **c** are conventionally chosen as short as the C-centring allows so that they need not be the shortest two vectors in their net plane and need not fulfil the inequalities.

Classifications

By means of the above-mentioned lattice properties, it is possible to classify lattices according to various criteria. Lattices can be subdivided with respect to their topological types of domains, resulting in two classes in two dimensions and five classes in three dimensions. They are called *Voronoi types*. If the classification involves topological *and symmetry* properties of the domains, 24 *Symmetrische* Sorten are obtained in three dimensions and 5 in two dimensions. Other classifications consider either the centring type or the point group of the lattice.

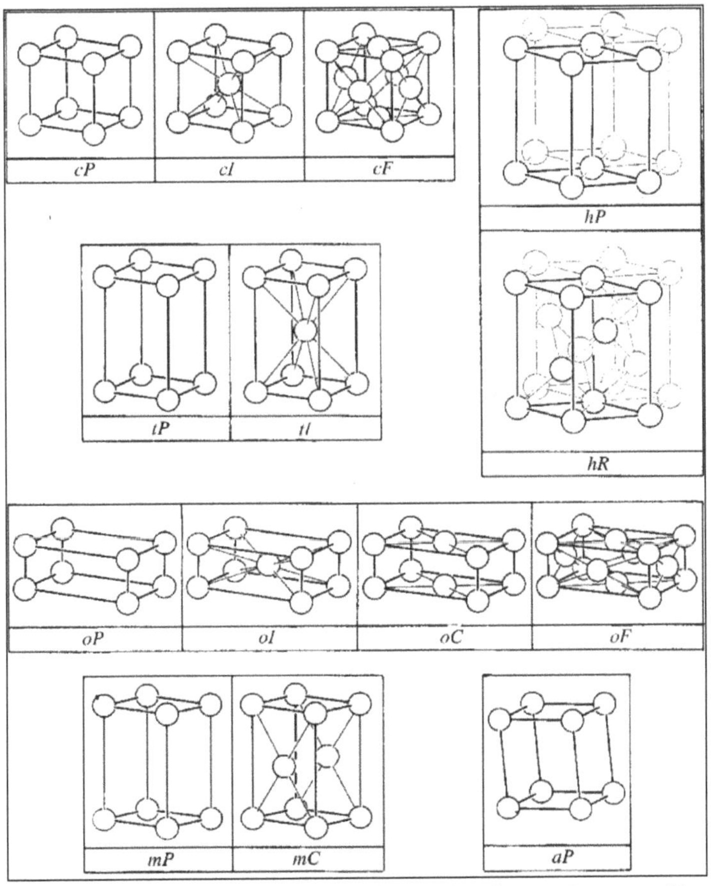

Figure: Conventional cells of the three-dimensional Bravais types of lattices.

The most important classification takes into account both the lattice point-group symmetry and the centring mode. The resulting classes are called *Bravais types of lattices* or, for short, Bravais lattices. Two lattices belong to the same Bravais type if and only if they coincide both in their point-group symmetry and in the centring mode of their conventional cells. The Bravais lattice characterizes the translational subgroup of a space group. The number of Bravais lattices is 1 in one dimension, 5 in two dimensions, 14 in three dimensions and 64 in four dimensions. The Bravais lattices may be derived by topological or algebraic procedures. It can be shown that 'all Bravais types of the same [crystal] family can be obtained from each other by the process of centring'. As a consequence, different Bravais types of the same crystal family differ in their centring mode. Thus, the Bravais types may be described by a lower-case letter designating the crystal family and an upper-case letter designating the centring mode. The relations between the point groups of the lattices and the crystal families are shown in Table. Since the hexagonal and rhombohedral Bravais types belong to the same crystal family, the rhombohedral lattice is described by hR, h indicating the family and R the centring type. This nomenclature was adopted for the 1969 reprint of International.

Permissions

All chapters in this book are published with permission under the Creative Commons Attribution Share Alike License or equivalent. Every chapter published in this book has been scrutinized by our experts. Their significance has been extensively debated. The topics covered herein carry significant information for a comprehensive understanding. They may even be implemented as practical applications or may be referred to as a beginning point for further studies.

We would like to thank the editorial team for lending their expertise to make the book truly unique. They have played a crucial role in the development of this book. Without their invaluable contributions this book wouldn't have been possible. They have made vital efforts to compile up to date information on the varied aspects of this subject to make this book a valuable addition to the collection of many professionals and students.

This book was conceptualized with the vision of imparting up-to-date and integrated information in this field. To ensure the same, a matchless editorial board was set up. Every individual on the board went through rigorous rounds of assessment to prove their worth. After which they invested a large part of their time researching and compiling the most relevant data for our readers.

The editorial board has been involved in producing this book since its inception. They have spent rigorous hours researching and exploring the diverse topics which have resulted in the successful publishing of this book. They have passed on their knowledge of decades through this book. To expedite this challenging task, the publisher supported the team at every step. A small team of assistant editors was also appointed to further simplify the editing procedure and attain best results for the readers.

Apart from the editorial board, the designing team has also invested a significant amount of their time in understanding the subject and creating the most relevant covers. They scrutinized every image to scout for the most suitable representation of the subject and create an appropriate cover for the book.

The publishing team has been an ardent support to the editorial, designing and production team. Their endless efforts to recruit the best for this project, has resulted in the accomplishment of this book. They are a veteran in the field of academics and their pool of knowledge is as vast as their experience in printing. Their expertise and guidance has proved useful at every step. Their uncompromising quality standards have made this book an exceptional effort. Their encouragement from time to time has been an inspiration for everyone.

The publisher and the editorial board hope that this book will prove to be a valuable piece of knowledge for students, practitioners and scholars across the globe.

Index

www.ingramcontent.com/pod-product-compliance
Lightning Source LLC
LaVergne TN
LVHW080721210926
848485LV00003B/500